# CHALLENGES TO THE ENLIGHTENMENT

We are concerned about the future of humankind. We do not believe that traditional religious and political orthodoxies are adequate to solve the problems that face us. We take as our primary goal the growth of human knowledge, the development of humane values based upon rationalistic foundations, with the defense of human freedom.

—From the founding statement of
The Academy of Humanism,
September 1983

# CHALLENGES TO THE ENLIGHTENMENT

## In Defense of Reason and Science

edited by Paul Kurtz
and Timothy J. Madigan
for
The Academy of Humanism

 **Prometheus Books**

59 John Glenn Drive
Buffalo, New York 14228-2197

Published 1994 by Prometheus Books

98 97 96 95 94     5 4 3 2 1

Library of Congress Cataloging-in-Publication Data

Challenges to the enlightenment : in defense of reason and science /
    the Academy of Humanism.
        p.    cm.
    Includes bibliographical references.
    ISBN 0-87975-869-4
    1. Science—Philosophy. 2. Science—Social Aspects. 3. Enlightenment.
4. Humanism. I. Academy of Humanism.
Q175.C4456    1994
501—dc20                                                    93-41680
                                                                CIP

Printed in the United States of America on acid-free paper.

# Contents

## Part II: Scientific Issues

## Part III: Social Issues

# Preface

The Academy of Humanism was founded in 1983 to honor prominent individuals who espouse the philosophy of humanism. There have been many different types of academies in the world, but the Academy of Humanism is unique. Its goals are to recognize distinguished humanists who, through their various efforts, have disseminated humanist ideals. Members of the Academy are nontheists who (1) are devoted to free inquiry in all fields of human endeavor; (2) are committed to a scientific outlook and the use of scientific methods in acquiring knowledge; and (3) wish to uphold humanist ethical values and principles. Membership in the Academy is limited to eighty persons. Humanist laureates are well-known in many fields and include Nobel Laureates, distinguished scientists, philosophers, authors, and artists.

Founded on the five hundredth anniversary of the Spanish Inquisition and the three hundred fiftieth anniversary of the trial of Galileo, the Academy's goals include furthering respect for human rights, freedom, and the dignity of the individual; tolerance and a commitment to social justice; the development of a universal perspective which transcends national, ethnic, religious, and racial barriers; and a belief in a free, open, and pluralistic democratic society.

These are essentially the values of the Enlightenment, with its dedication to reason as the best means of dealing with the myriad problems that face humankind. The Enlightenment was an intellectual

9

movement which began in seventeenth-century Europe and espoused an optimistic project: an end to human ignorance and the slavish adherence to ancient texts and dogmas; the annihilation of inequality between the sexes; and the advocacy of political rights for all citizens.

Traditional challenges to the Enlightenment came from orthodox religious camps, which looked with suspicion upon efforts to achieve advances without supernatural support. Recently, however, the ideals of the Enlightenment project have been assailed by a new set of critics, known under the loose label of "postmodernists." These critics have raised serious objections to the ability of human reason to grasp objective reality, the reliability and efficiency of the scientific method, and the "tyranny of democratic elites."

*Challenges to the Enlightenment* grew out of a meeting held by the Academy of Humanism in 1992, in Utrecht, the Netherlands. The meeting was called to respond to both the premodern and postmodern critiques of the Enlightenment project. The volume is divided into three parts: philosophical issues, scientific issues, and social issues. Several of the papers delivered at the conference are here published for the first time. Others, which were previously published, are reprinted with permission. Although diverse in subject matter, the following papers address a unifying theme: the need to keep alive the ideals of the Enlightenment in a world that is becoming increasingly alienated from a scientific and secular perspective. The Academy of Humanism is in the forefront of this important endeavor.

Paul Kurtz

# Part I

# Philosophical Issues

# 1

# Toward a New Enlightenment*

## Paul Kurtz

We are witness to a period in history when there is a great deal of disenchantment with the Enlightenment expressed by a wide range of intellectuals. These critics are often described as "antihumanist," though some who identify themselves as "humanists" also participate in the attacks on the Enlightenment as being at best "outdated" and "illusory," or at worst "repressive" and "evil." The term *postmodernist* characterizes a diverse set of writers who are critical of "modernity," which is equated in part with the Enlightenment. The term *Enlightenment* has been used to refer to certain intellectual trends in the seventeenth and especially the eighteenth centuries in Western society. It began perhaps with Descartes, Bacon, and Locke, who proposed the use of reason, experience, or science as a universal method for obtaining knowledge, and it culminated with the French *les philosophes*—Voltaire, Diderot, Condorcet, d'Holbach, and the Encyclopedists.

Enlightenment writers exuded great optimism about the potential of science and reason in unlocking the secrets of nature and understanding human nature, and in applying this knowledge for the betterment of

---

*Originally published in *Free Inquiry* (Winter 1992/93): 35–37. Reprinted with permission.

the human condition. They had faith in the power of education to transform society. They were foes of religious superstition and mythology, and they criticized the influence of clericalism. Some were atheists, but most were deists. They maintained that there were universal ethical norms that transcended cultural relativity. They believed in the ideals of liberty, equality, the secular society, and democracy. Many attacked the *anciens regimes,* defended industry and commerce, and wished to use technology to improve social conditions. They thought that nature in general and human nature in particular were basically good. They believed that human progress could be attained and happiness widely distributed for the greater good.

The term *modernity* refers to the fact that there was confidence in the ability of men and women to control their destiny. They believed that human beings were free, autonomous, and rational agents, and were responsible in some measure for their future. They were convinced that, by means of science, objective knowledge was possible about nature and our place within it. By enlightened understanding and action, they thought that we could improve human life and create a more just and beneficent society.

Although there was some disagreement about one or another aspect of the above outlook, Enlightenment ideals inspired many subsequent thinkers, from Kant, Goethe, Bentham, and Mill, to Marx, Darwin, and Freud, and they helped to fuel the rapid development of the sciences and economic, social, and political change. This is not without recognizing the Romantic protest of the nineteenth century—the view that reason was not enough and that we needed to supplement thought with passion, the will, the arts, spirituality, and other dimensions of human experience. In a real sense, the Enlightenment realized many of the ideals of the Renaissance and the humanism of ancient Greece and Rome.

If we look back two centuries, we cannot but be impressed by the significant impact of the Enlightenment: first, in the continued growth of the scientific revolution and the expansion of knowledge; second, in the unparallelled effect of technology in transforming, taming, and conquering the planet, reducing human suffering, poverty, and disease, and contributing to human well-being and happiness; third, in the impressive advance in universal education, literacy, and learning—once considered the privilege of only the upper class—and now viewed virtually everywhere as the universal right of children of all classes, rich and poor alike; fourth, in the advancement of the democratic revolution, which has swept far beyond France, England, and America (whose foundations, incidentally, were made possible by the disciples of the Enlightenment: Jefferson, Madison, Franklin, Paine, etc.) to all continents, so that the defense of liberty, equality, and human rights is now accepted by the world community.

What then is the cause for despair? Why should so many intellectuals condemn the Enlightenment, modernity, reason, and science, even freedom and democracy—and single out humanism as the arch-foe? If at the end of the nineteenth century, Nietzsche could proclaim the "death of God," as we approach the end of the twentieth century, so many declare "the death of man." Theodor Adorno perhaps expresses a pervasive attitude about humanism and its optimistic faith in man when he states that "after Auschwitz" we cannot write hymns "to the grandeur of man."[1]

At any one time in history, there are no doubt many cultural streams competing for ascendency. For every generalization offered, a counter-generalization can be found. The truth is that both humanism and the forces of antihumanism exist side by side in contemporary society. What are some of the later forms of antihumanism? Clearly, the hope of the Enlightenment that superstition would be vanquished by education and science has not been fully realized, as neo-fundamentalism everywhere abounds. Traditional Catholicism, Protestant fundamentalism, Orthodox Judaism, a revitalized Islam, and Hinduism all deny that humans are capable of mastering their own fate, and they demand submission to their mythologies of salvation. Aleksandr Solzhenitsyn's attack upon secular humanism in his Harvard University address in 1978[2] and his call for a reawakening of a new nationalistic spiritualism is symptomatic of the times. In recent years we have observed the reemergence of forces that many moderns had thought were long since vanquished: intense nationalistic, ethnic, religious, and racial passions; and the outburst of multicultural chauvinisms of the crudest sort (from Croatia and Serbia to Armenia and Slovakia, from Northern Ireland and the West Bank to Quebec and Native American tribalism). All call for a renewal of ethnic loyalties. We may ask: Is nationalism more powerful than rationalism; are blood and soil more enduring than universal ideals?

I do not intend to focus on this problem here, however, but rather on a far more sophisticated critique of the Enlightenment and humanism that in one sense seems to imply not simply a retreat from reason, but a collapse into subjectivism and despair and a loss of confidence that we can control the future—a new form of *nihilism*.

Undoubtedly, one of the key ingredients in this pessimism about the humanist agenda is the apparent death of Marxism, an ideology that had attracted so many intellectuals. It has been pointed out by the critics of humanism that both Marx and Robespierre were committed to the Enlightenment. They believed in reason and progress, and both ideologies ended in terror and/or the totalitarian gulag. To save the revolution, any means could be employed. Marxist-Leninists attacked democratic freedoms and even Marxist humanism, as only a reflection of "bourgeois liberalism." Today Marxists, if there are any left, condemn Stalinism and defend human rights. Yet the dreams of so many communist and socialist idealists were shattered by the totalitarian experience. It is difficult to stimulate the conviction that science, secularism, or humanism are viable social goals. Who could have imagined a decade ago that the theology of *laissez faire* capitalism would be heralded in Moscow, East Berlin, Warsaw, and Budapest, and that the quest for ancient tribal loyalties and the old traditions would resurface with such a vengeance?

Humanists need to point out that they were among the first to criticize the use of terror and to defend the open society, and that they have always placed the freedom of the individual at the top of the list of primary values.

Yet the humanist emphasis on human freedom is today attacked from still another quarter. A powerful indictment is now heard from the latter-day disciples of Martin Heidegger. Indeed, his philosophical outlook is fashionable among faculties in the elite universities in the Western world. French postmodernists Jacques Derrida, Jacques Lacan, Michel Foucault, and Jean-François Lyotard are in vogue. Interestingly, they not only reject the Enlightenment project but explicitly deny any number of humanist premises: that human beings are capable of free and autonomous choice; that they can be rational and responsible; that universal ethical norms can be discovered; that meta-narratives of emancipation can or should be achieved; that the ideals of liberal democracy and of human rights have genuine authenticity. Drawing upon Heidegger's philosophy, they likewise deplore the growth of technology. They maintain that language is a veil masking Being, that every text should be deconstructed, and that objective scientific knowledge is a myth.

Derrida is heralded as a virtual prophet of postmodernism. Recently he was granted an honorary degree from Cambridge University, although not without vigorous protest from many faculty.[3] Derrida is influenced by Heidegger's famous "Letter on Humanism,"[4] in which the latter rejects Sartre's view that existentialism is a humanism and Sartre's defense of radical human freedom and autonomy.[5] What are we to make of this assault on humanism by the French Heideggerians? We may ask, What is the ethical and social praxis to which their philosophical principles lead?

Of considerable embarrassment to postmodern Heideggerians is the publication of Victor Farias's book *Heidegger and Nazism*,[6] in which he points out that Heidegger joined the Nazi Party in 1933, when he became rector of Freiburg University, and he retained his membership in the party until 1945. Heidegger assumed the chair once held by Husserl at Freiburg. According to Hannah Arendt, he was "the uncrowned king of the empire of thought." As rector, he welcomed the National Socialists. He subsequently resigned, but in 1935 he reaffirmed the "inner truth and greatness of National Socialism." Whether he was anti-Semitic is open to dispute. He declined to direct the dissertations of Jewish students,[7] though he thought that Jewish traits were "cultural" rather than "biological." In 1966, in an interview in *Der Spiegel,* published posthumously at his insistence, he claimed that his adherence to the Nazis came from the brief conviction that they were the only hope of the German nation and that they could best grasp "the problem of technology." In 1947 Herbert Marcuse, his former student, implored Heidegger to publicly renounce his identification with Nazism. Heidegger's reply was to equate the forcible transplanting of East Germans with the annihilation of Jews. And later he compared the new agricultural technologies with "the manufacturing of corpses in the gas chambers and extermination camps."[8] All of this should be viewed in the light of Heidegger's repudiation in his writings of democratic liberalism and his call for a new philosophy.[9] Some defenders of Heidegger maintain that his philosophical contributions should be separated from his political convictions and/or his naïveté in this regard. But how should we interpret the writings of a great philosopher, if not by examining in part the consequences of his philosophy in ethical and social practice, given the fact that his writings reflect not simply ontological or epistemological reflections but ethical pronouncements throughout? How shall we view his rejection of the ethics of humanism?

Incredibly, Philippe Lacoue-Labarthe, a disciple of Derrida, has sought to defend Heidegger by maintaining that "Nazism is a humanism." Why? Because "it rests on a determination of *Humanitas,*" he says, "which is, in its eyes, more powerful, i.e., more effective, than any other."[10] Like other forms of humanism, it too, he argues, attempts to impose human categories on Being.

But this ploy is a scandalous distortion. For what is distinctive of humanism is its uncompromising defense of human dignity and value, and its commitment to human freedom, the equality and worth of every person—ideals betrayed by Heidegger and the National Socialists.

A telling response to postmodernist Heideggerians and a defense of humanism appear in a series of books published by two French philosophers, Luc Ferry and Alain Renaut.[11] They argue that Heidegger's critique of humanism has been thoroughly discredited.

"We have reached the conviction," they say, "that his [Heidegger's] indictment of modern times and humanism, which he saw as going back to Descartes and the philosophy of the Enlightenment, could at the very best lead to a radical criticism of every feature of the democratic world: the world of technology and mass culture, of course, but also the world of human rights. . . ."

"It is impossible," however, they say, "to return, after Marx, Nietzsche, Freud, and Heidegger to the idea that man is the master of the totality of his actions and ideas." Today we know about the illusions and dangers implicit in denying the role of the unconscious. But they maintain that the key question is to rethink "the question of the subject,"[12] and also to develop a new form of humanism, which would include a defense of human freedom, democracy, and human rights.

To totally reject "modernity" and the entire set of ideals of the Enlightenment is sheer folly. For there is much that we are indebted to that we cannot abandon. I submit that we need to use the best of the ideals of the Age of Reason, but to adapt them to the contemporary world. The key contributions of modernity are still meaningful, but perhaps it is only as a "post-postmodernity," or a new humanist Renaissance. We need a *re*construction of human knowledge and values, not a *de*construction, a *re*vision rather than a derision of human potentialities. We need to reaffirm some optimism about the human condition in the place of the regnant pessimism.

I say this at a time when social, political, and economic change worldwide is so rapid that it is often difficult for anyone to predict or even recommend with confidence what will or indeed should happen in the future. At the present moment the Cold War is over and the threat of a nuclear holocaust—at least temporarily—has diminished. The great European colonial empires have disappeared. Pax Britannica and the glory of France have been replaced by Pax Americana, the sole superpower, which is being challenged by Japan, a new Germany, and a new Europe. There has been a collapse of American optimism, and all that its neo-conservative leaders have had to offer was a return to the ancient Judaeo-Christian faith and opposition to secular humanism.

The overwhelming challenge in the contemporary world is the continued growth of technology (as both Heidegger and John Dewey recognized) and the failure of humankind to know how to deal with it. Shall we assume the nihilistic posture of postmodernists and recoil in horror at science and technology, or recognize its potentialities for good or evil and attempt to use its fruits wisely for the benefit of humankind? The crisis that we face is the disparity between the new powers that we possess (the knowledge explosion: for example, biotechnology, space technology, the computer sciences) and our prevailing values that often are based on ancient myths and dogmatic religious systems. Here humanism can provide an authentic alternative, indeed, the most viable option, for it is the *only* meaningful outlook or life-stance (eupraxophy) on the world scene that consciously and forthrightly defends the scientific outlook and its methods for coping with the world.

Let me briefly outline what I think are some of the key features of a new post-postmodern neo-humanism and its relevance to the future.

First, the age of science continues to proceed, and in three senses. Humanism is its major philosophic or eupraxophic expression.

In the continuing controversy of what is reality, humanists maintain that the sciences probably best describe what we encounter in nature and account for or explain how and why it is occurring. Thus, humanism crystalizes a *cosmic outlook:* an evolutionary universe in which the concepts, hypotheses, and theories on the frontiers of the natural, biological, and social sciences are taken seriously, rather than those of theology or poetry. Metaphysical speculation or appeals to revelation, faith, intuition, or emotion cannot be a substitute for experimental inquiry and theoretical confirmation. What we surely lack today is *sophia,* or wisdom, integrating the body of knowledge. Unfortunately, science is divided into narrow specialties, and one can be competent in one field and not in others. The lone trailblazing scientist has been replaced by researchers employed by governments or multinational corporations who work for purposes of power or profit. We need to recapture the scientific outlook and develop some *sophia* about its broader implications for ourselves and the public at large. In the mass media, what we often hear are sensationalized interpretations—such as that the big bang proves the existence of God, or that near-death research proves the existence of the afterlife. Part of the scientific outlook is its skepticism about the traditional theistic outlook. We need to engage in the continuing critique of biblical claims by scholarly, linguistic, archaeological, and historical inquirers and to defend the naturalistic alternative.

There is still an exciting dynamic aspect to the expanding frontiers of science, and some place for optimism that research problems, though difficult, can be resolved by introducing new hypotheses and theories and testing them experimentally.

It is clear, however, that science is not simply a body of fixed principles, or part of an encyclopedia of knowledge, but that it is defined by its *methods of inquiry* in understanding nature and human behavior and in coping with problems that we encounter. The use of reason does not mean abstract reason or absolute truth. Science is tentative, fallible, probabilistic. Although it is able to make steady progress, there are continuing changes and revisions. There are, however, I submit, some *objective methods* that we use, and these are continuously tested in the real world by their consequences. Hence science is *not* simply one myth among others (as some of the postmodernists maintain), nor is it simply a question of a dominant historical paradigm that decides between scientific theories. Although clearly influenced by the socio-cultural context, its methods of inquiry are tested by their demonstrated effectiveness in comparison with other methods.

It is the application of science to technology that exemplifies its major impact. Industrial society was based on the technology of heavy industry; the society of the future on information technology. Would the philosophers or poets who meditate about the universe and rail against technology wish to give up antibiotics, modern surgery, pace-makers, artificial respirators, or anesthetics, which have extended the human life-span and have reduced pain and suffering, or the green revolution in agricultural technology, which has provided enriched standards of nutrition and reduced hunger and poverty? Do they wish to give up their flush toilets (which an engineering colleague at my university considers the most important invention of the nineteenth century), refrigeration, running tap water, electricity, stereophonic equipment, or printing presses, which print their books and magazines? Do they wish to return to the technology of simple peasants and scribes?

Clearly there are abuses of uncontrolled technology, and ecological and biodiversity destruction. But it is not technology per se that is at fault, but its unwise use.

Second, the central question concerns our *ethical values.* Here I submit that the values of humanism are meaningful. They already have a wide acceptance in the world and are on the cutting edge of social change. We need to be clear about what our *eupraxia* (i.e., our ethical practice and life stance) entails.

Our first focus here is on *freedom of thought and conscience* and *free inquiry.* Granted that this is limited or conditioned by the social context, yet we need to reaffirm the vitality of the inquisitive and probing mind. Are the readers of Heidegger or Marx free to accept or reject their arguments, or is free, autonomous, or rational inquiry an illusion?

Even to pose this question presupposes some capacity for free inquiry.

Humanists defend the *right to privacy,* self-determination, moral freedom, the right of the individual to make his or her own choices about love and sex, family and friends, career and profession, tastes and desires, medical care, life and death decisions—consonant, of course, with the rights of others.

I submit that there are *objective ethical standards* implicit in humanism. I call these the "common moral decencies" and the "values of excellence." Ethics need not degenerate into subjective taste or caprice. Reason, however, must be cojoined with passion, and cognition can modify and reconstruct emotion. Although we recognize a wide diversity in values, there are general ethical norms that apply to humankind in general. I would defend a neo-Kantian and modified utilitarian theory. One cannot separate means from ends. This was a major failure of Marxist theory. That is why we need an ethics of both principles and values.

Third, humanism offers a significant *social theory.* Again, this is making great headway in the world. It is the philosophy of *democracy and the open society,* political and economic, of tolerance and respect for differences. And this is intrinsically related to human rights. It transcends cultural relativity and provides general normative principles of behavior.

Of great challenge here is the emergence of a *planetary* or *global ethic,* which takes the viewpoint of humankind as a whole. This entails the need to develop an environmental ethics. In this regard, we will need in the future to deal with the unregulated power of multinational corporations, the disparity between rich and poor countries, the urgency of population control, the warfare between ethnic and nationalistic tribal groups. Humanism offers a universalistic outlook based on a common science and common values. Recognizing cultural diversity, it is the party of humankind, for it is concerned with the world community beyond ethnic, racial, or religious factions.

Fourth, we need to make it clear that humanism provides a response to the central question of *the meaning of life.* If the orthodox drama of divine salvation and immortality of the soul has no evidential merit, what is the alternative? The humanist life stance offers a viable option: the good life of creative fulfillment, happiness and exuberance for the individual person. Human existence need not be bereft of ideals to live by, significant plans and projects to strive mightily to attain. The great problem for humanism in the future, however, is to elevate the level of tastes and the qualities of appreciation, enrich cultural expression, and enhance educational opportunities for all.

Fifth, and most important, is the genuine possibility of some realistic *optimism* about human potentialities and the human prospect. In this postmodern age we have been thoroughly disabused of any claims to achieving unlimited progress. There is no end to history, there are only new beginnings; and every day is a challenge to us to create our own world, and to strive for a better future. If we cannot build a utopian society, at least we can ameliorate the human condition. But if we are to do so, it is not by a retreat into pessimistic despair, encouraged by the Cassandras in our midst, nor fearful anxiety. It requires the willingness to express the key humanist virtues of cognition and courage, mixed with compassion, and a resolve to enter into the world and change it for the good. If we are to do so, we will need a *re*-enchantment with the ideals of humanism, a *re*-enlightenment. We need a new Enlightenment. For those who say it is not possible, I say it is, and indeed, the humanist stream of culture is making headway in spite of its critics.

History is not fixed. There are no inevitable laws of social development that we discover. What will occur depends on us. Whether the twenty-first century and beyond will be the humanist century depends in part upon Lady Luck and chance, the contingent and unexpected, but it also depends upon our efforts and what we do. Given these considerations, I submit, humanism still has a bright prospect.

# Notes

1. Theodor Adorno, quoted in Luc Ferry and Alain Renaut, *French Philosophy of the Sixties: An Essay on Anti-Humanism* (Amherst, Mass.: University of Massachusetts Press, 1990), p. xxix. Published originally in France as *La Pensee 68: Essai sur l'anti-humanisme contemporain* (Paris: Gallimard, 1985).

2. See excerpts from Aleksandr Solzhenitsyn's talk and Sidney Hook's

article, "Solzhenitsyn and Secular Humanism: A Response," *The Humanist* (November/December 1978).

3. I am rather amused by this since I think that I deserve some credit, or blame, for first introducing Derrida to the American public. I organized a French/American philosophers' conference at the State University of New York in 1968, at which time he read his influential paper, "The Ends of Man." This was published in *Philosophy and Phenomenological Research* and in a book I edited, Paul Kurtz, ed., *Language and Human Nature* (St. Louis: Wm. H. Green, 1971).

4. Martin Heidegger, "Letter on Humanism," in *Basic Writings,* ed. D. F. Krell (New York: Harper and Row, 1977). There are humanistic themes in Heidegger's *Being and Time* (1927), but he disavows humanism in his later "Letter on Humanism."

5. Jean-Paul Sartre, *Existentialism and Humanism* (1946), translated into English by P. Mairet (London, 1948).

6. Victor Farias, *Heidegger and Nazism,* trans. by Joseph Margolis and Tom Rockmore (Philadelphia: Temple University Press, 1989). Published in France as *Heidegger et la Nazisme* in 1987.

7. A former colleague of mine, Marvin Farber, was a student of Husserl at Freiburg and knew Heidegger. Farber was responsible for bringing phenomenology to the United States, and founding the journal *Philosophy and Phenomenological Research.* He was convinced that Heidegger had compromised all his philosophical principles because of his Nazi associations.

8. I was a graduate student of Herbert Marcuse in 1948 at Columbia University and remember the development of his own philosophical position and the influence of Marx, Hegel, Freud, and Heidegger on his work.

9. An excellent review of Farias's book is by Thelma Lavine (*The Washington Post,* 1989). She clearly points out the complicity of Heidegger and its relevance to his philosophical contributions.

10. Philippe Lacoue-Labarthe, *La fiction du politique: Heidegger, l'art et la politique* (Paris: Christian Bourgois, 1987), p. 81.

11. See especially Luc Ferry and Alain Renaut, *French Philosophy of the Sixties: An Essay on Anti-Humanism;* Luc Ferry and Alain Renaut, *Heidegger and Modernity* (Chicago: University of Chicago Press, 1990), published originally in France as *Heidegger et les modernes* (Paris: Grasset & Fasquelle, 1988).

12. Luc Ferry and Alain Renaut, *French Philosophy of the Sixties,* p. xvi, Preface to the English Translation.

# 2

# Counter-Enlightenment in Contemporary Social Studies

## Mario Bunge

## Introduction

Modernity was born in the seventeenth century and it reached adulthood in the eighteenth. This was the *siècle des lumières* or period of the Enlightenment. There may always have been enlightened people and obscurantists. But it was only in the eighteenth century that a systematic and concerted effort was made to constitute an enlightened ideology guiding a powerful cultural and political movement that was to achieve a profound and progressive social transformation. That was the century of the American and French revolutions, of the beginning of secularism and liberalism, and of the deification of reason and the exaltation of science, technology and industry. It was an age of progress and optimism. It was the second Renaissance. Reaction and pessimism were to come later.

We have come a long way from the Enlightenment. We have made enormous progress but we cannot take progress for granted any longer: instead, we must work for it and fight its enemies. Moreover, we must admit that much of the progress we have made has had its lead lining. In the face of nuclear arms, overpopulation, environmental degradation, the rapid depletion of non-renewable natural resources, the persistence of colonial exploitation and racism, the impoverishment of the Third World, the concentration of wealth in a few hands in a few countries, the mass production of cultural junk, and the recent revival of fascism and nationalism, it would be foolish to believe in automatic progress on all fronts.

However, it would be even more foolish, and indeed suicidal, to believe that we are condemned to regression and eventual extinction. In particular, it would be foolish and suicidal to despair of our ability to redesign the future and secure the survival of humankind with the help of science, technology, and a humanistic and unselfish morality. Yet in our "postmodern" era it has become fashionable to surrender to such despair and moreover to blame reason, science and technology— rather than our wrong values and our political and business leaders— for our current predicament.

The new Romanticism is fashionable partly because the illusions and promises of my generation have not been fulfilled, and partly because it is the easy thing to believe, because inaction is easier than action, and because irrationalism is favored by the most reactionary forces, which thrive on our ignorance and unwillingness to tackle social issues in a rational and realistic manner. As Isaac Asimov has said, it is far easier and less dangerous to reject science and technology than to revolt against the social order: the former only takes ignorance and it does not put one's life or freedom on the line.

The retreat from the Enlightenment is fashionable but far from original. In fact, the Counter-Enlightenment followed right on the heels of the Enlightenment, it was revived less than a century ago, it triumphed briefly with Nazism, and it is fashionable again under the headings of Counterculture and Postmodernism. Two centuries ago Herder exclaimed "I am not here to think, but to be, feel, live." Between 1925 and 1945 the Nazis sought to remythologize modern society by replacing the cult of universal reason with that of blood and soil. Nowadays millions of young people listen to Skid Row, one of the heavy-metal groups, whose motto is "If you think, you stink."

We shall examine the "postmodern" reaction in the field of social studies. However, it will help us understand this reaction if we start by placing it in its historical context.

## Peculiarities of the Enlightenment

The catchwords of the Enlightenment were *nature* and *humankind, reason* and *science, liberty* and *equality, happiness* and *utility, work* and *progress.* Reason was placed at the very center of this constellation: if only men were to think and act rationally, the rest would follow. The thinkers of the *siècle des lumières,* who were often called *philosophes,* saw themselves as the bearers of the torch of reason and explorers of the world previously shrouded in the darkness inherited from the Dark Ages.

Suffice it to remember the *Encyclopédie* (begun in 1751), the Declaration of American Independence (1776), and the Déclaration des droits de l'homme et du citoyen (1789), as well as such diverse thinkers as Locke and Hume, Voltaire and Condillac, Montesquieu and Condorcet, Diderot and d'Alembert, Buffon and Lavoisier, Helvétius and d'Holbach, Quesnay and Smith, Beccaria and Bentham and—though intellectually minor, closer to us—Ben Franklin and Tom Paine.

The Enlightenment was a whole new ideology: a world view, a value system and a political agenda. It was the first comprehensive modern ideology, and one wedged between another two comprehensive ideologies: Thomism and Marxism—which in some ways was a continuation of, and in others a reaction against, the Enlightenment (more on this in the next section).

The Enlightenment ideology may be compressed into the following ten principles:

1. Trust in reason—culminating during the French Revolution in the pathetic worship of the goddess Reason.

2. Rejection of myth, superstition, and generally groundless belief or dogma.

3. Free inquiry and secularism, as well as encouragement of deism (in contrast to theism), agnosticism, or even atheism.

4. Naturalism (as opposed to supernaturalism), in particular materialism.

5. Scientism: adoption of the scientific approach to the study of society as well as nature.

6. Utilitarianism (as against both religious morality and secular deontologism).

7. Respect for craftsmanship, enthusiasm for industry, and reverence for the machine.

8. Modernism and progressivism: contempt for the past (except for Classical Antiquity), criticism of present shortcomings and vices, and trust in the future.

9. individualism together with libertarianism, egalitarianism (to some degree or other), and political democracy—not yet for women or slaves.

10. Universalism or cosmopolitanism—e.g., human rights and education for all "free men."

Some of these principles are part of a world view, others are value judgments, and still others are parts of a program to change society. Although the Enlightenment ideology was a coherent system, some people adopted some of its components while rejecting others. For example, the enlightened European despots and some of the American patricians, both North and South, adopted the philosophical but not the political components of the Enlightenment ideology. In particular, they were against obscurantism but not for social emancipation. On the other hand the Romantic philosophers and political thinkers rejected all ten principles: they were reactionary all the way.

## The First Counter-Enlightenment

Romanticism, a powerful cultural movement that spanned a quarter of a century on each side of 1800, is usually presented as a reaction against the Enlightenment and the French Revolution, as well as against the Napoleonic invasions. So it was with regard to philosophy and politics. But Romanticism was not just a reaction: there was also Romantic art. And this art was extraordinarily original and rich in literature and music. Furthermore, it was hardly touched by philosophy or politics.

We shall therefore distinguish three strains in Romanticism: the artistic, the philosophical, and the political. Romantic art was philosophically and politically neutral, philosophical Romanticism was obscurantist, and political Romanticism was conservative or reactionary. A few examples will make this point clear. To begin with Rousseau—

often regarded as the first Romantic—was politically progressive but philosophically obscurantist, for exalting feeling over reason and for holding that science had a pernicious influence on society. On the other hand Heine, Beethoven, and Shelley were artistically romantic but they sided with the world view and the political philosophy of the Enlightenment. Goethe and Blake were politically progressive but anti-intellectualist and in particular they disliked modern science. Closer to us, most of the ideologists of the New Right are rationalists, whereas the New Left (or what is left of it), as well as the Feminist and the Green movements, teem with irrationalists—in particular, enemies of science and technology.

Having drawn a distinction between artistic, intellectual, and political Romanticism, we shall henceforth concentrate on the middle strain, since it is the one most relevant to our topic. Intellectual romanticism, in contradistinction to artistic romanticism, is a reaction against the philosophy and the value system of the Enlightenment. It is either irrationalist (in particular intuitionist) or idealist; it is antiempiricist (in particular idealist or apriorist), and it is antiscientific and technophobic. In short, intellectual Romanticism is traditionalist or antimodernist.

The core of intellectual Romanticism was the idealist philosophy of Fichte, Schelling, Hegel, Herder, and Schopenhauer. Though different, all five men were idealists, they opposed the burgeoning science of their time, and attempted to counter the *Entzauberung* or demythologization process that Max Weber, echoing Auguste Comte, saw as the trademark of modernity.

The Romantic philosophers identified logic with ontology, a confusion that gave them leave to speculate freely about reality: they often mistook fiction for fact. They wished to replace natural science with natural philosophy, and social science with social (in particular legal) philosophy. They believed all things to be organic wholes opaque to analysis. They opposed conceptual and empirical analysis, claiming that dissection kills.

In the field of ethics and political philosophy the Romantics proposed replacing bourgeois individualism with holism or collectivism. (On the other hand the Romantic poets and musicians were individualists.) Last, but not least, the prose of the Romantic philosophers was remarkably obscure and stuffy. They are the ones who invented the trick of passing off nonsense for profundity. This trick was perfected in our century by the phenomenologists, existentialists, hermeneuticists, and deconstructionists.

As with other intellectual currents, in this case there were a number of border cases. Immanuel Kant was one of them. He was as enlightened as anyone could be in the most backward of Prussian provinces. He respected mathematics and science although he hardly understood them, and he was a universalist, cosmopolitan, and pacifist. But, because of his idealism and intuitionism, his insistence on the limits of reason, and his dogma that psychology and the study of society could never become sciences, Kant prepared the way for the onset of German Romantic philosophy.

Another case of contamination of the Enlightenment by Romanticism was Marxism. Marx and Engels thought of themselves as the continuers of the Enlightenment and the French Revolution. This they were in some important respects: they were serious, important social scientists; they pushed liberalism toward the left—with characteristically romantic enthusiasm; they were materialists on the whole; and they wrote clearly except about dialectics.

However, Marx and Engels learned from Hegel a few lessons that vitiated their whole system. One was dialectical ontology, according to which everything is a unity of opposites, and that opposition, "contradiction" or conflict is the source of all change. A second Hegelian inheritance was the equation of logic with ontology. A third was the belief that, in order to establish the truth of a statement of fact, it suffices to invoke a few favorable examples, without bothering to look for exceptions or counterexamples. A fourth Hegelian inheritance was holism or collectivism, i.e., the thesis that the whole determines the part—e.g., that the individual is at the mercy of social systems and historical forces. A fifth was an ethical consequence of holism, namely that individual rights are as nothing compared with the duty to submit to "historical necessity."

I submit that this legacy of Hegel was definitely obscurantist and it favored the solidification of Marxism into a dogma disconnected from "bourgeois" philosophy and social science. Notwithstanding these important legacies of Hegelianism, original Marxism was on the whole anti-Romantic. (On the other hand neo-Marxism, in particular Stalinist Marxism and structuralist Marxism, often accentuated the Romantic streak in Marx and Engels, in particular the dialectical mumbo-jumbo and the contempt for empirical testing.) So, there is still hope of salvaging some of the ingredients of Marxism—precisely those that harmonize with the Enlightenment.

## The Second and Third Romantic Waves

The second Romantic wave came roughly one century after the birth of the first. It was initiated by Wilhelm Dilthey—the father of philosophical hermeneutics—and Friedrich Nietzsche, who was to become the favorite philosopher of Hitler and Heidegger. Other outstanding members of the second wave were Heinrich Rickert—Weber's philosophical mentor—Henri Bergson, Hans Vaihinger, and even the latter William James—rightly called a Romantic utilitarian. The neo-Hegelians Croce and Gentile (Mussolini's co-author and Minister of Education) joined later on. This was a motley collection: some of them were intuitionists, others idealists, others radical skeptics, and still others pragmatists. But they all shared a mistrust of reason, in particular formal logic and mathematics, and in general a lack of confidence in science. None of them cared for empirical tests, and some of them did not even find use for the concept of truth. Most of them were antidemocratic as well.

The third Romantic wave, which I shall call neo-Romanticism, overlapped partially with the second. It began with Husserl's phenomenology, was followed by Heidegger's existentialism, and culminated in "post-modernism" and the contemporary antiscience and antitechnology movement. Some of the best known names in this movement are Edmund Husserl and Martin Heidegger, Oswald Spengler and Jacques Ellul, Georg Lukács and Louis Althusser, Albert Camus and Jean-Paul Sartre, Karl Jaspers and Hans-Georg Gadamer, Michel Foucault and Jacques Derrida, Thomas Kuhn and Paul Feyerabend, Clifford Geertz and Harold Garfinkel, Barry Barnes and Bruno Latour.

Though quite different from one another, these authors share most or all of the following five typically Romantic traits. These are (1) the mistrust of reason and, in particular, of logic and science; (2) subjectivism, or the doctrine that the world is our representation; (3) relativism, or the negation of the existence of universal truths; (4) an obsession with symbol, myth, metaphor, and rhetoric; and (5) pessimism, or the denial of the possibility of progress—particularly in matters of knowledge.

Most of the neo-Romantics write inexact and often impenetrable prose—another Romantic trait. (Remember Nietzsche's contempt for "the offensive simplicity of the style" of John Stuart Mill or, even better, Heidegger's meaningless and therefore untranslatable strings of words.) Furthermore, they are not interested in moral problems and consequently they have no moral philosophy to offer other than either moral individualism (egoism) or moral collectivism (conformism). Finally, not daring to be thought of as antimodern—that is, reactionary—some of these writers call themselves *postmodern*—an oxymoron, as befits irrationalism.[1]

Up until the 1960s the influence of neo-Romanticism was confined to Germany, France, and their spheres of cultural influence. The major intellectual currents in the Anglo-American world and its sphere of cultural influence had been proscientific, protechnological, and antiobscurantist. In particular, the academic community worldwide had on the whole been committed to critical thinking and empirical evidence, as well as to liberal politics. The heritage of the Enlightenment was taken for granted. Moreover, we were supposed to have entered the Space and Computer Age. Ours was supposed to be the postindustrial society, run by knowledge rather than by labor. Finally, the continuous economic growth of the industrialized countries between ca. 1950 and 1970 suggested that "we" (that is, one-fifth of humankind) had reached the age of plenty and high quality.

The combination of affluence and war changed all that in the United States. Middle class affluence eroded the work ethic and facilitated hedonism and egoism. The Me Generation ushered in. The Vietnam War and later on the acceleration of the arms race gobbled up many of the funds earmarked for the liberal social programs. Large sectors of the population became alienated or disaffected. Many students and young academics began to question the received value system and the concomitant ideology—especially when facing either the military draft or unemployment. They became receptive to ideas and lifestyles alternative to the so-called American Way of Life. Among them were drugs and irrationalism—two perfect mates.

From about the mid 1960s on German and French neo-Romanticism was imported into the Anglo-American academic world. It became influential not just among literary critics and philosophers but also among students of society, particularly anthropologists and sociologists. The close traditional ties between this new wave of obscurantism and totalitarian politics were severed or covered up for purposes of consumption in the new market. So much so that many have mistaken the revolt against reason for a rebellion against the "establishment"—as if logic and basic science were ideologically tainted, when in fact irrationalism is an effective intellectual tool of oppression, for blunting critical thinking and valuation.

Let us now examine the influence of neo-Romanticism on social studies in recent times.

## Critical Theory and Phenomenological Sociology

Let us take a quick look at four neo-Romantic schools of social studies: critical theory, phenomenological sociology, ethnomethodology, and the constructivist-relativist sociology of science. All four reject the scientific approach, regard science proper as one more myth or ideology, and are characterized by uncritical thinking and hermetic language. Let us start by examining the first two schools.

Critical theory, also called the Frankfurt school, is the variety of humanistic (or philosophical or armchair) sociology characterized by a critique of capitalism and all known types of social order; the denial of the distinction between science and ideology; the denunciation of science and technology as handmaidens of capitalism; the rejection of positivism, rationalism, and scientism; repetitive invocations to Hegel, Marx, and Freud; the demand that social science become a tool for social change; and long-winded, heavy, and opaque prose devoid of analysis and argument, of figures and formulas.[2] Much the same holds for the French structuralists and post-structuralists, in particular Louis Althusser and Michel Foucault.

We shall not touch on the Frankfurt-school critique of capitalism, which derives essentially from Marxism—though without whatever empirical support the latter may muster. We shall concentrate instead on some of the philosophical components of critical theory. Firstly, like Marxism, critical theory adopts uncritically Hegelian dialectics: it does not doubt that everything is a unity of opposites, and that the source of every change is "contradiction" or the struggle of opposites. It does not elucidate the terms "opposite" and "struggle," and it overlooks all the exceptions to the "law" that "contradiction" (opposition) is the engine of change. In particular, it overlooks production, trade, cooperation, demographic change, and technological innovation, none of which can easily be construed as a struggle of opposites—except in the sense that the new sometimes opposes the old.

Secondly, critical theorists reject rationality, believing that it is the supreme tool of the domination of man. Thirdly, they embrace un-critically psychoanalysis, with all of its wild speculations, thus showing that they are gullible and incapable of distinguishing pseudoscience from science. Fourthly, they mistake science for technology, and consequently they are incapable of understanding that basic science cannot be used directly as an agent of social change, and that it is needed not only to understand society but also to redesign it. Fifthly, they reject the distinction between science and ideology. Indeed, the central points of critical theory are that (a) science = technology = capitalist ideology; and (b) sociologists are necessarily committed to either the conservation or the alteration of society—hence the adjective "critical." Moreover, they hold that science is not ethically neutral or even epistemologically objective. They brand as "positivists" those who claim that it is possible to gain objective knowledge of social facts. In other words, they equate science and positivism, thus rendering the latter an undeserved service.

Between the two world wars the Frankfurt school tackled, admittedly in a rather literary way, some real social issues: the evils of unregulated capitalism, the crimes of Nazism and Stalinism, and the impotence of liberalism. It performed a useful if modest ideological function even while disparaging the objective study of those social movements. Since then critical theorists have continued to criticize capitalism and totalitarianism in imprecise general terms but they have turned increasingly to idealistic (in particular neo-Kantian) philosophy, and they have had nothing precise to say about the most pressing social issues in the post-war period, such as overpopulation, environmental degradation, the arms race, the North-South chasm, "stagflation," and ethnic conflicts. In particular, Habermas has gradually shifted from an interest in social conflict and oppression to "communicative action," and has openly consorted with the hermeneuticists.

Moreover, by rejecting the scientific approach to social issues the critical theorists block the understanding of such issues as well as any attempts to tackle them rationally and therefore effectively. In this way, despite its revolutionary rhetoric, critical theory has become a conservative force, a sort of academically respectable safety valve for nonconformists, and one more variety of obscurantism. To top it all, for all their publications over seven decades, critical theorists have not elucidated any key sociological concepts or proposed any original and testable hypotheses—much less, theories proper. In sum, critical theory is neither.

Our second example is phenomenological sociology.[3] This is the conservative counterpart of critical theory. It is characterized by spiritualism and subjectivism, as well as by individualism (both ontological and methodological) and conservatism—ethical and political. The first two features are obvious: according to phenomenology social reality is a construction, not a given, for all facts are "meaningful" and thus are to be "interpreted." Consequently everything social is spiritual and subjective or at most intersubjective, rather than material and objective.

The ontological individualism of phenomenology derives from its subjectivism. Because individuals are said to "interpret" themselves and others, without ever facing any brute social facts, the task of the sociologist is to grasp "subjective meaning structures" rather than to construct and test sociological models of social systems. In particular, he must study the *Lebenswelt* (everyday life) of individuals, staying away from large social issues such as gender and race discrimination, social conflict and political oppression, militarism and colonialism. The phenomenologist can capture directly the objects of his study because they are ordinary.

Moreover, he is graced by the "vision of essences," a special insight which allows him instant grasping. Hence he can dispense with tedious fact-finding, arguments, and empirical tests. In short, phenomenological sociology is admittedly unscientific.

The ethics and politics of phenomenology are clear: Far from being subjected to social constraints, the individual is autonomous because he constructs social reality. Hence there is no reason to bother about emancipation. The sociologist ought to be interested only in social order, because men crave "meaning" and order. He ought to shun conflict and, in general, social issues. The social summum bonum is stability, with its accompanying rigid order and certainty, not social progress with the concomitant disorder, risk, and uncertainty. In short, phenomenology is ethically and politically conservative. Consequently it is not a guide for any social policy other than "law and order."

## Ethnomethodology

Our third example of the neo-Romantic reaction against social science is ethnomethodology, an offspring of the union of phenomenology with symbolic interactionism.[4] Ethnomethodologists practice what phenomenologist sociologists preach: They observe at first hand trivial events in the Lebenswelt or everyday life, focusing on communication and staying clear of any important social activities and issues. They engage in participant observation and shun experiment, which they disapprove of on philosophical grounds.

Lacking theories of their own, most ethnomethodologists adopt the murky pronouncements of phenomenology and even existentialism—two notorious radical enemies of science. Obviously, an antiscientific philosophy could hardly inspire scientific research. Mercifully the ethnomethodologists make no use of these doctrines in their empirical work. In fact, when in the field they behave as positivists even while vehemently denouncing positivism, inasmuch as they spend most of their time collecting data which they cannot assimilate for want of theory.

In fact, the ethnomethodologists audiotape and videotape "the detailed and observable practices which make the incarnate production of ordinary social facts, for example, order of service in a queue, sequential order in a conversation, and the order of skillfully embodied improvised conduct."5 English translation: The ethnomethodologists record observable ordinary life events.

The data thus collected are the audible or visible traces left by people who presumably behave purposefully and intelligently. This is the only clue the ethnomethodologists can go by, for, lacking a theory, they cannot explain what makes people tick. The ethnomethodologist's practice does not differ from that of the empiricist and, in particular, the behaviorist. In short, he behaves like a positivist even while engaging in positivism-bashing—which is actually an attack upon the scientific approach.

Only the ethnomethodologists' convoluted lingo suggests intimate contact with their philosophical mentors. Example: Garfinkel6 starts one of his books stating that ethnomethodology "recommends" that "the activities whereby members [?] produce and manage settings [?] of organized everyday affairs are identical with members' procedures for making those settings 'account-able'[?]. The 'reflexive'[?] or 'incarnate'[?] character of accounting practices [?] and accounts makes up the crux of that recommendation." Why such opaque prose to describe ordinary accounts of everyday life?

This is not to deny the value of observing everyday life occurrences, such as casual encounters and conversations—the favorite material of ethnomethodologists. The observation of ordinary life, a common practice of anthropologists, yields raw material for the scientist to process in the light of hypotheses and with a view to coming up with new hypotheses. But that material is of limited use unless the subjects are placed in their social systems, for only this may tell us why they behave as they do. In fact, ethnomethodology deliberately overlooks the macrosocial context and consequently all large social issues. This, combined with the absence of tests of the proposed "interpretations" (hypotheses) and the lack of theory, explains the poverty of findings of ethnomethodology.

Let us see how ethnomethodology views scientific research.7 Its findings are essentially two. One is that "something more" is involved in scientific research than what can be formulated in even the most detailed of instruction manuals. This "something more" is the set of tacit assumptions and of bits and pieces of know-how (procedural knowledge), both of them well known to psychologists, philosophers, and engineers. The other "finding" is that, no matter how elementary a scientific experiment may be, it cannot be performed without a modicum of theory—this being why a partially paralyzed chemistry student could do his lab exercises with the help of an able-bodied ethnomethodology student, largely ignorant of chemistry, who acted as the former's hand. But did we not know this all along, at least those of us who have had a scientific training and did not fall for the most crass form of empiricism? And if the ethnomethodologists understand that there is no genuine science without theory, why don't they produce one?

Being radical individualists, and focusing on such everyday life practices as conversation, ritual, and entertainment (e.g., Balinese cockfight), the ethnomethodologists openly admit their indifference to the problems of social structure and, indeed, to all social issues—as recommended by Husserl and Schutz. By concentrating on individuals (or at most pairs of individuals) and overlooking all important social activities and social systems, they fail to come to grips with social reality. Why should people who study only trivial facts, have no theory to speak of, and write in obscure language be counted as social scientists?

## Constructivist-Relativist Sociology of Science

The sociology of science is, of course, the study of the interactions between scientific communities and society at large. The scientific study of such interactions came to maturity right after World War II around Robert K. Merton.[8] The neo-Romantics attacked Merton's school, accusing it of being naive in holding that scientists search disinterestedly for objective truths, that they abide by a moral code, and that they share freely their findings. This attack was based on a systematic confusion of basic science with technology, and of the latter with industry and politics. Its aim was was not to purify the scientific enterprise but to denigrate it.

The pseudophilosophical weapons employed by the anti-Mertonian sociologists of science were found in the armamentarium of the third

Romantic wave. In fact, the practitioners of the new sociology of science have revived the old idealist thesis that the subject (or else the group) constructs reality instead of exploring and discovering it: they are constructivists. In other words, they do not distinguish facts from our statements and theories about them. Consequently they do not need the concept of objective truth: each individual (or group) would construct his own truths as he constructs his (or its) own world. In short, the school in question is relativist as well as constructivist. This suggests that science is one more myth, ideology, or tool of political power.[9]

Here is a sample from one of the best known publications in the Neo-Romantic sociology of science, Latour and Woolgar's *Laboratory Life*.[10] They write about "the social construction of scientific facts," and claim that mere conversational exchanges and "negotiations" among researchers can "create or destroy fact." They assert that " 'out-there-ness' [the external world] is the *consequence* of scientific work rather than its *cause*."[11] Consequently "a scientist's activity is directed, not toward 'reality,' but toward . . . operations on statements."[12]

Moreover, according to this school, all scientific activity boils down to making inscriptions, reading them, or quarreling over them. The world is a book, and the task of the scientist is that of a symbol interpreter or hermeneuticist. Since the inscriptions are public objects, the sociologist of science needs no special training to interpret them. Thus, without knowing any physics or mathematics, Latour[13] claimed to analyze the special theory of relativity and announced a momentous discovery. Indeed, according to him Einstein was wrong in believing that his theory had anything to do with the electrodynamics of moving bodies—the title of his 1905 paper. Latour assures us that the theory in question is about certain human activities and, more particularly, that it is a set of "new instructions for bringing back long-distance scientific travelers."[14]

The members of the neo-Romantic sociology of science never produce any evidence for their quaint assertions: they are dogmatic like their philosophical mentors. On the other hand there is plenty of evidence against their views. For example, we all know that the world existed long before there was anyone around capable of making any inscriptions. Only madmen and philosophers can afford to believe that they construct the world. All normal people distinguish between objective facts and our ideas about them, and cherish objective truth. We all know that scientists and others discover independently existing things in addition to inventing ideas. In particular, we all know that scientists and others make observations and measurements to find out something about the world, to activate theories, or to test hypotheses about reality.

The constructivist-relativist sociology and philosophy of science is not just blatantly false and barren. It is also dangerous because it propagates the irrationalist myth that science is just inscription-making, and thus no different from fiction writing or pseudoscience. Because it denigrates the epistemic value and the morality of science, it diverts some young people from the study of science. Indeed, the current world-wide decline in the student enrollment in science courses is due in part to the neo-Romantic attack upon science. Why make the effort to study science if it is no different from fantastic literature or religion and, moreover, it is a tool of oppression?[15]

## Conclusion

The Enlightenment gave us most of the basic values of contemporary civilized life, such as trust in reason, the passion of free inquiry, and egalitarianism. Of course the Enlightenment did not do everything for us: no single social movement can do everything for posterity—there is no end to history. For instance, the Enlightenment did not foresee the abuses of industrialization, it failed to stress the need for peace, it exaggerated individualism, it extolled competition at the expense of cooperation, it did not go far enough in social reform, and it did not care much for women or for the underdeveloped peoples. However, the Enlightenment did perfect, praise, and diffuse the main conceptual and moral tools for advancing beyond itself.

On the other hand the faithful of the Counter-Enlightenment would have us set the clock backward rather than tackle the current issues and try to go forward. They are barbarians intent on destroying modern culture while continuing to enjoy its technological spin-offs. Although they constitute a motley crowd, basically they only differ amongst themselves by the intensity of their hatred of reason and science—which they conveniently dub "positivism." Not surprisingly, they have produced no remarkable findings, not even new interesting errors the denials of which would constitute valuable truths.

Ordinary mistakes and scientific errors can be detected and corrected in the light of reason or experience. But when reason and experience are written off, such correction becomes impossible, errors are perpetuated, and cheap nonsense and shallow rhetoric replace the laborious search for truth. Worse, when obscurantism is in the ascendancy freedom and progress are at risk. When this happens the so-called eggheads are mobbed by skinheads, whose muddled brains control booted legs anxious to trample on the Enlightenment legacy.

## Acknowledgments

I am grateful to David Blitz, Bernard Dubrovsky, and Jacques Herman for enlightening discussions. The research leading to this paper has been supported by the Social Sciences and Humanities Research Council of Canada.

## Notes

1. See, e.g., M. Featherstone, ed. "Special issue on Postmodernism," *Theory, Culture & Society* 5 (1988): 195–576; and D. Harvey, *The Condition of Postmodernity* (Oxford: Basil Blackwell, 1989).

2. See, e.g., A. Arato and E. Gebhardt, eds., *The Essential Frankfurt School Reader* (Oxford: Basil Blackwell, 1978).

3. A. Schutz, *The Phenomenology of the Social World* (Evanston, Ill.: Northwestern University Press, 1967); and P. Berger and T. Luckmann, *The Social Construction of Reality* (London: Allen Lane, 1967).

4. H. Garfinkel, *Studies in Ethnomethodology* (Englewood Cliffs, N.J.: Prentice-Hall, 1967); and C. Geertz, *The Interpretation of Cultures* (New York: Basic Books, 1973).

5. M. Lynch, E. Livingston, and H. Garfinkel, "Temporal Order in Laboratory Work." In *Science Observed: Perspectives on the Social Study of*

*Science,* eds. K. Knorr-Cetina and M. Mulkay (London: Sage, 1983), pp. 205–238.

6. H. Garfinkel, *Studies in Ethnomethodology,* p. 1.

7. See Lynch, Livingston, and Garfinkel, "Temporal Order in Laboratory Work," for a summary.

8. R. K. Merton, *The Sociology of Science: Theoretical and Empirical Investigations* (Chicago: University of Chicago Press, 1973).

9. For representative selections, see B. Barnes, ed., *Sociology of Science: Selected Readings* (London: Penguin, 1977); and K. Knorr-Cetina and M. Mulkay, eds., *Science Observed.*

10. B. Latour and S. Woolgar, *Laboratory Life* (London and Beverly Hills, Calif.: Sage, 1979; rev. ed.: Princeton, N.J.: Princeton University Press, 1986).

11. Ibid., rev. ed. (1986), p. 182.

12. Ibid. (1979), p. 237.

13. B. Latour, "A Relativistic Account of Einstein's Relativity." *Social Studies of Science* 18 (1988): 3–44.

14. Ibid., p. 23.

15. For further criticisms, see M. Bunge, "A Critical Examination of the New Sociology of Science, Part 1." *Philosophy of the Social Sciences* 21 (1991): 524–60; "A Critical Examination of the New Sociology of Science, Part 2." *Philosophy of the Social Sciences* 22 (1992): 46–76.

# 3

# Hostility to Science

## John Passmore

The enemies of science march in many different armies, sound very different trumpets, wave very different banners, perhaps religious, perhaps metaphysical, perhaps environmentalist, perhaps feminist. Sometimes they claim to be fighting in the interests of humanity, the humanities, or even, I fear, humanism. I shall have to concentrate on a very few central metaphysical and moral issues relating for the most part to science itself rather than its technological progeny.

Hostility to science, on such grounds, dates back to the very beginnings of scientific speculation in Ancient Greece. One finds it clearly delineated in Aristophanes' comedy *The Clouds.* True enough, Aristophanes' aim is indiscriminately wide. He was a master of that propaganda device which puts all your enemies in the one basket. So he ascribes to Socrates, against whom his satire is particularly directed, not only that critical attitude to the Olympian religion with which Socrates is certainly associated but the teachings of the Sophists, which Socrates detested, and the new biological and physical theories which, according to Plato's *Phaedo,* he regarded as being totally insufficient. (Insufficient because they did not contribute to what Socrates saw as the only really worthwhile question—why it is best for things to be as they are, what purpose their being so serves.)

Nevertheless, Aristophanes' conservative instincts were not leading him wholly astray. Whatever Socrates' own intentions, the critical methods he had introduced, the new social doctrines of the Sophists, the new physics and the new biology all had this in common: they threatened the central myths of Athenian society. Socrates looked critically at the traditional stories about the gods, condemning them because they made of the gods beings who were all too human in their passions; the Sophists looked skeptically at the view that Athenian laws originated in semi-divine legislators rather than arising out of particular social circumstances; the natural scientists no longer permitted the Athenians to see in such natural phenomena as rain, lightning, thunder the work of divine anger and divine beneficence. It was all, Aristophanes depicts them as saying, a matter of clouds; striking against one another, clouds produced thunder and burst into rain.

One need not take Aristophanes too seriously as a science-reporter. It is the spirit of these doctrines that matters. Here Aristophanes got it dead right. A new attitude to the world was emerging—one in which the human being was, in important respects, much less at home, a world in which very many things happened not as a result of human or divine, human-like, passions or to fulfill human or human-like purposes but as a result of the operations of humanly indifferent natural processes. If Socrates resisted that movement of ideas, defending a more sophisticated version of the old teleology, he nevertheless contributed to the destruction of the old gods, gods one might properly fear but could nonetheless joke about, and the eventual substitution of what Pascal was to call the 'God of the philosophers,' with no trace of the laughing Bacchus or the amorous Jove, an abstract Being.

This was all two thousand five hundred years ago and you may think it absurd to evoke the names of Socrates and Aristophanes in these so modern days. But if human beings are now very different in respect to the kind of knowledge and the kind of power which, as a species, they possess, they have in many quite fundamental respects changed very little.

Some of us, no doubt, have learnt to live in a world in which things happen, for the most part, in a manner which is quite independent of anyone's passions or intentions, divine or human. We may even prefer such a world to one which is thought of as being under providential control, finding it easier to bear the horrors of the world if we can think of them as having risen out of the chances of natural selection, appreciating the more deeply what is good in the world just because things could so easily have been still worse. But many people, whether scientists or non-scientists, clever or stupid, ignorant or well-informed, cannot accept such a world-view, any more than the conservative Aristophanes could. Their hostility to it surfaces as prayers for rain, as creationist attacks on evolution, as stories about visitors from other planets, as occultism, as that pseudo-science which is more dangerous to science than antiscience, or as antiscience itself.

How is this metaphysical form of hostility to science to be met? Some defenders of science would, of course, reply that metaphysical fears of science have no ground, that science does not *necessitate* the existence of the kind of world I have just described. Indeed, we are witnessing a revival of attempts like those of Jeans and Eddington in the nineteen-thirties to deduce theological conclusions from physics and cosmology. But it is enough for many antiscientists that science has at least helped to generate and sustain it. So much is scarcely deniable. My own defense would be very different: that such a vision of the world in no way compels us to surrender what is genuinely valuable and indeed encourages us to cherish it the more. That science somehow subtracts from the value of generosity, compassion, tolerance, from the beauty whether of a sunset or a Titian, or that on the other side it makes cruelty, bigotry, the ruthless exercise of power, the devastations of industry less evil—this, certainly, is an illusion. On the contrary, the world-view I described brings forcibly home to us the rarity and precariousness of what is valuable, whether at the level of ordinary human decencies or the highest human achievements. Only when science itself falls victim to ancient human aspirations for a perfected human being in a perfected nature, when it forgets evil, is there good reason to fear it. To that point I shall later return.

There is a strong tendency, however, among human beings to believe that the existence of evil and misery must be the result of the operations, preferably a single operation, of some human or human-like person.

The most extraordinary example of this is the Adam and Eve story where a wrongful act—theologians disagree about the precise kind of act it was—on the part of Eve and then of Adam is held to be responsible for the fact that human beings do not live a wholly paradisiacal life. In classical mythology, it was another single action, the opening of Pandora's box, which released into the world all its evils and miseries. Pandora was created by the gods as a very special kind of being— etymologically her name means the all-gifted one—in revenge for a prior act on the part of Prometheus, his stealing fire from the gods and giving it to human beings. There are two features of this story which are relevant to my theme. The first is that once again the miseries of the world are blamed on a single act, the opening of Pandora's box, the second is that ultimately the source of all evils and miseries is taken to be the most central of all technologies—the use of fire—and the box was borne by a woman who was 'all-gifted,' as if all human gifts are danger- ous. If, in the Genesis story, the knowledge provided to Adam and Eve by their eating of the apple is of good and and evil rather than of science and technology, there is still there, too, the suggestion that "ignorance is bliss."

So although for my part I find the picture of the world which I sketched above quite congenial, it has generally, over human history, been replaced by a quite different view, in which some human or divine action, or a mixture of the two, is responsible both for the very existence of moral evil and human misery in the world and for particular evils and particular miseries. The human action is often ascribed to a group. Quite recently I heard someone say: "So many things have gone wrong at once; there must be some one group which is responsible." Scientists are sometimes nominated as that one group, or at best they are resented for offering nonpersonal explanations of phenomena which the anti- scientists want to ascribe to human or divine causes or for complicating life by offering multicausal explanations. For these explanations make it impossible to suppose that all human ills can be cured by some single act, whether by seeking divine intervention or by getting rid of the group whose actions are the single cause of whatever goes wrong, a role which Jews have often played, or more recently Communists.

I do not, of course, want to defend what is sometimes taken to be the scientific view, that all moral evil and human misery has its source in impersonal forces. Individuals, including ourselves, can sometimes rightly be blamed for acting as we do and so can groups, terrorists for example. That their actions have causes does not make them any less blameworthy. We can so act, on the positive side, as to reduce misery and the incidence of moral evil. All that I am trying to do is to offer a degree of explanation why some people react with hostility to the world-view which science has helped to establish, a world-view which looks, in many cases, to impersonal causes in areas where human beings have historically looked to personal or divine causes and therefore often, not always, puts us in a situation where there is no one to blame, no one to propitate.

Aristophanes attacked science from yet another, very different, point of view. Science, his comedy suggests, concerns itself with what are, humanly speaking, trivialities. That line of attack persisted for a very long time. It is central in Dean Swift's parody of the Royal Society, in Dickens' parody of the British Association for the Advancement of Science; one still meets it in the writings of newspaper columnists, complaining about the kind of research on which public funds are spent.

Without describing the necessary technology, Aristophanes introduces Socrates as in a basket, high above the air. There, we are told, "he could more clearly penetrate the things of heaven," could "traverse the air and contemplate the sun." "Had I remained on the ground," Socrates tells us, "I should have discovered nothing." The earth's atmosphere would have interfered with his thinking. This introduction to Socrates, foreshadowing modern space observation by satellites, has in Aristophanes a double satirical function. It ascribes to Socrates that dreaded Greek vice *hubris,* an attempt by human beings to lift themselves above the human state, to become godlike. But it also suggests that by raising himself above the earth Socrates was leaving the place where the really important problems are.

The same charge that science sets aside the important for the inconsequential is implicit in Aristophanes' earlier description of two pieces of biological investigation. It is no accident that the first is about fleas, the second about gnats, those perennial symbols of insignificance. (Remember the phrases: "not worth a flea," "straining at a gnat"). Socrates is represented as asking the question: "How many times the length of its legs does a flea jump?" and answering it experimentally. He first calculated the size of the flea's legs by dipping them in wax and then he removed the resulting wax boot; he thus measured the length of the jump. In the second example Socrates is asked whether a gnat buzzed through its proboscis or its rear. He answers that "the gut of the gnat was narrow, and in passing through this tiny passage, the air was driven with force toward the breech; then after this slender passage it encountered the rump, which was distended like a trumpet, and there it resounded sonorously." To which Strepsiades, that still-familiar figure, a merchant in difficulties who has decided it was time for him to try the new technologies, is made to reply. "So the rear of a gnat is a trumpet! Oh, what a splendid discovery. Thrice happy Socrates! It ought to be easy enough to succeed in a law-suit, knowing so much about the gut of a gnat!" Aristophanes' irony is unmistakable. "To succeed in a law-suit"—that is the world of practical affairs, here and now as in litigious Athens. How the audience would have roared at this juxtaposition of such trivialities as a knowledge of gnats and fleas with such serious matters as success in a law-suit!

If these two accusations—that science is metaphysically and morally dangerous and that it concerns itself with trivial matters—ran for a long time in tandem, the second of them at least, or so it might be thought, could now be maintained only by those who are either merely prejudiced or totally ignorant of what science has achieved. The practical usefulness of having made a careful study of fleas and mosquitoes was apparent once scientists recognized their importance as disease vectors; the complaints so often heard in the earlier nineteenth century that scientists were wasting their time playing around with electricity now seem ridiculous. Even the United States, so hard to convince until the Second World War that science, outside such applied fields as medicine and agriculture, was any more worthy of public support than the arts— that is, not at all—has experienced a conversion on a grand scale.

Yet there are signs that the position is altering, that the older attitude is reasserting itself, if in a form modified by experience. There is a growing suspicion, if I read the signs aright, of unsubstantiated claims for technological spin-offs of benefit to human beings; the "you never

know" technique is wearing thin. Voices are to be heard pointing out
that there seems to be no relation between a country's capacity to win
Nobel Prizes and its technological success; that Japan seems largely
to rely on what used, not so very long ago, to be called "yankee know-
how" at a time when the United States was still a scientifically backward
country but industrially advanced, reaching that point by buying or
appropriating such science as its innovations needed. Such voices sometimes suggest, too, that new technological innovations are likely to come
not from new science but from practical inventiveness.

At the practical level, this attitude shows itself in anxious University
pronouncements that they are no longer "Ivory Towers," in the tendency
of research councils to set up projects and then look for scientists to
work at them rather than to leave judgments about what is important
to scientists themselves. One sees a growing realization, at least, that
what a scientist counts as "important" and what as "trivial" may not
at all accord with public judgments, employing different criteria. So
far Aristophanes' parody is still relevant, even if for his "fleas and gnats"
we have to substitute, let us say, nuclear particles.

Faced with the demand that they concentrate their attention on
publicly important issues the scientists have to stand up and be counted,
have to be prepared to say, publicly and loudly, "What we count as
important is a better understanding of the world; if you want science
to flourish you must let us decide what are the most likely routes to
that understanding; past experience suggests that a better understanding
often increases human powers but we can never guarantee in any particular case that this will be so and that is not our governing purpose."

That is easier for the philosopher to recommend than for the scientist
to do; the practice of science is expensive, philosophy cheap. Money
is most easily obtained for research which promises practical gains. But
the public is by no means indifferent to the importance of pure science
as a source of understanding rather than of power; much of the most
expensive work now going on, in astronomy for example, is not likely
to issue in practically important technological innovations. It could do
so, but I do not think that public readiness to pay for it rests on such
a foundation. And something similar is true, at the opposite extreme,
of much practically trivial work in biology.

In the long run scientists can do themselves a great deal of harm by pretending that their work is likely to have the sort of consequences a Strepsiades would accept as being important when the probability that it will do so is miniscule. I am not suggesting that scientists should simply shrug their shoulders when it is suggested to them that a particular line of work, if it turns out to be successful, would be of great social importance. Still less am I suggesting that they should, in a form of intellectual snobbery, look down on those who depart from the more straightforward paths of scientific success to tackle obdurate messy problems of great practical importance, where the chance of failure is high and even success is unlikely to win the highest scientific honors. All I am saying is that they should not engage in false pretenses, however powerful may be the temptation to do so, in the new scientific world of a never-ceasing quest for research funds.

The moral standing of science is by no means unimportant to it. It is given special privileges; large sums of money are entrusted to it; it is called upon to give authoritative advice in a great range of areas. When he linked the physical scientists with the Greek Sophists, Aristophanes was suggesting that they cannot in fact be morally trusted. This view the nineteenth century disputed, if with dissentients, scientists were set up as moral paragons in respect to their willingness to set aside financial rewards, power, public honors in the quest for truth, in respect to honesty, truthfulness, impartiality, devotion, cooperativeness, tolerance of criticism, generosity of spirit, and, furthermore, as people trained in these virtues by the very character of science as an institution. But the older Aristophanic view now has its advocates.

Any institution for which such claims are made must expect to be greeted with a degree of cynicism and resentment; there is no mistaking the gusto with which in some medieval depictions of the Last Judgment cardinals are depicted as being sent to Hell. No one who has ever been at all close to science will doubt that if science strengthens certain virtues it, like every other human institution, also generates particular vices: sometimes an unpleasant degree of arrogance, often joined with complacency, the taking of credit by senior scientists for work in which they have scarcely participated, a lack of scruples in experimenting, an undue preoccupation with honors and awards. A *New Scientist* article once suggested indeed that "scientific establishments are notorious for bickering, infighting, jealousies and conspiratorial maneuvering."

I do not think that the friends of science need to be at all startled by these judgments or shocked to the core by the sharply competitive dealings made public in *The Double Helix*. That scientists are human too, as given to jealousy, paranoia, power-seeking as the rest of us, is scarcely surprising. It is not surprising, either, that there are characteristic scientific vices just as there are characteristic vices generated in the world of art, of the church, of the army, of politics. None of this is inconsistent with the view that science is nevertheless a school where certain virtues are particularly cultivated.

Much more serious are those moral weaknesses which find expression in the falsification of the results of experiments. These properly created a public scandal—comparable to the discovery that doctors have abused their special relations with patients or teachers with their pupils. In each such case there is a failure on the part of individuals which weakens the trust on which the entire enterprise rests. That the scandal in such instances is so great is, however, a testimony to the general health of science, however fatal it may be to the doctrine that science contains so many correcting mechanisms that faked results are bound immediately to be exposed.

At a more fundamental level there are those who would have us believe that the supposed virtues of scientists are completely fraudulent. There is, on their view, no such thing as objectivity. Scientists are social hirelings, seeking greater power for themselves under the pretense that what they are seeking is understanding. Science, as they see it, is simply a typical capitalist institution which has been more successful than most in concealing its hypocrisy.

The attack on the concept of objectivity, now quite fashionable in considerable areas of sociology and even the philosophy of science, cuts very deeply. I have considered it elsewhere in a discussion of objectivity in history; what I say there can, I think, be applied to the natural sciences, with certain modifications and extensions. The essential point is this: one can easily define objectivity in so extreme a way—just as in the case of such concepts as "certainty," "knowledge," "truth"—that nothing satisfies the requirements one has laid down. Calvin substantially did this with "human virtue." But on any more sensible standards, science encourages and develops objectivity. One cannot, in general, deduce from what scientists take to be reliable information that they come from a particular country or to what political or religious system they adhere, let alone their personal character. There are some marginal points, no doubt, at which one can make such deductions—scientists, as I said, are only human—but in no other major sphere of activity, excluding everyday commonplace judgments, is there quite so striking a divorce between ideological bent and systematic beliefs.

Other challenges have still to be met. The first is that objectivity, as the scientist understands it, is not a virtue but a vice, leading scientists to set aside human feelings that they ought not to set aside. It is in fact, on this view, a mask for callousness; with "objectivity" as their excuse, scientists engage in experiments which are sometimes cruel, whether to human beings or to animals, or which do not respect personal rights and personal dignity, or which are immoral, or are dangerous in their consequences. I do not see how it can be denied that all these things have in fact happened—especially when, as in Nazi Germany, scientists were temporarily freed from certain social constraints—but not only there. Of course, scientists can sometimes reply that they do not share the moral outlook of those who condemn them, who have the wrong ordering of values. So as against certain Vatican pronouncements biologists might argue that they see no harm in masturbation and great good in making it possible, by artificial insemination, for women to be able to have the babies they want to have.

Nevertheless, it is now generally admitted that experimenters are sometimes morally unscrupulous, that the enthusiasm which is essential to science can degenerate into a fanaticism for which nothing counts, no moral restraints, except the making of a particular discovery. Even if it appears that recombinant biology does not present the dangers that were at one time feared, it is significant that the demand for a moratorium came from biologists themselves, as it is also significant that ethics committees are being set up in hospitals. Scientists, then, are not prepared wholly to trust their fellow-scientists not to engage in experiments which, even by the most liberal standards, are morally impermissible whether in themselves or in their consequences. The literary image of the mad scientist, relating especially, since the time of *Franken-stein,* to biology, expresses a fear which cannot be set aside as wholly without justification. One can only partly assuage that fear by pointing out that the need for controls is now more widely accepted than it once was, if often very reluctantly. A physiologist who tells us that he used to advise his students: "Never say 'no' to an experiment" also tells us that he has now abandoned that dictum. But other physiologists, one must grant, may be less ready to change their minds.

Another objection to my modified defense of the morality of science is that I am now hopelessly out of date. What I said about the scientist as a model of devotion, disinterestedness, honesty, and the like, it is then admitted, might once have in some measure been true, in what is now often referred to as "the heroic age" of science. But that, it is claimed, was one of the casualties of the Second World War. Science on this view is no longer a paradigm of free and open discussion in a world in which so many scientists work in defense or corporation laboratories and, even in universities, keep their work secret in the hope of patenting it or setting up corporations of their own. Neither does such a scientist set aside the pursuit of power and worldly gain in his quest for truth.

One familiar reply to such critics, with a broad application in the antiscience debate, is that they confuse scientists with technologists. Technologists, even when remarkable, as many of them are, for their intellectual capacities, their imagination, their aesthetic feelings, have never pretended that they are, or could be, anything but the servants either of industry or of the state, working to specifications which are laid down for them, at best within broad limits.

Certainly when science is now described as "dangerous" it is often just in virtue of its technological applications, whether what is in question is nuclear warfare, pollution from chemical factories, or new biological techniques of reproduction. But at this point, scientists are in a certain difficulty. Needing funds as they do, they are not inclined to fall back on the old Cambridge, England, attitude: "Thank God our science is practically useless," even when that is in fact the case. Scientists would like to claim responsibility as the intellectual progenitor of such technology as is generally admired, while yet disclaiming responsibility for such of it as is generally regarded as, or turns out to be, threatening.

This is a clearly untenable position. But to ask how far science can properly be held responsible for science-based technology as a whole, as distinct from what I have elsewhere called "practitioner inventions," is to face a number of problems. One problem is whether, except at the extreme ends, this distinction is now a real one, i.e., whether—unless one is prepared to redefine science in such lofty terms that a very considerable number of those, inside as well as outside universities, whom the world commonly counts as such will have to be reclassified—there is not a large grey area which, like technology in general, pursues a practical end, but unlike most technology, extends scientific knowledge in the course of doing so. Another problem is how far the responsibility of science for technology extends.

Clearly, there could be no science-based technology were there no science. But the gap between the science and the technology is sometimes considerable, sometimes inconsiderable. One can distinguish three classes of cases. In the first case, the scientists did not, and could not have, anticipated the technology which relies upon it; its application depends on the unpredictable ingenuity of technologists, the discovery of new materials, other scientific discoveries, new demands. In the second case, the scientists could have anticipated how their work would have been used had they taken the trouble to engage in a certain measure of reflection. In the third case, they see quite plainly how their discoveries will be used and may even, as in the nuclear bomb case, suggest a use for them. In the first instance, while antiscientists can still argue that it would be better if science did not exist since we cannot predict what technologies it will make possible, they cannot reasonably blame the individual scientists for the consequences of their acts; in the third case, they obviously can do so—if we hand a gun to someone who we know intends to use it to shoot, it is no excuse that we did not ourselves pull the trigger. In the second case, the scientists, in my judgment, can equally be blamed if the outcome is forseeably calamitous. They have a duty of care, of circumspection.

On this moral issue, then, I must make the following admissions: first, that what the nineteenth century took to be the most morally attractive characteristics of scientists are now less evident than they once were; second, the character of modern science, its intense competitiveness, its constant search for funds, its use of large research teams, has accentuated some of the less attractive moral features of science; third, many people who are ordinarily accounted scientists are professionally engaged in work which, in the eyes of many, sometimes including myself, is morally reprehensible, and this is partly because they do not stop and consider the broader consequences of what they are doing.

Yet after all these admissions, I would still stand by the Enlightenment doctrine—there are virtues which science particularly generates and sustains. This is so even if, in our essentially commercial society, the scientist is as likely as, let us say, the artist, to be tempted by the attractions of money and power. I want to conclude, however, by taking up this last question—the temptations of power.

In the nineteen-thirties I was disturbed by the attitude of a great many scientists, particularly Cambridge scientists, to Stalin's Russia and, more generally, to such then fashionable concepts as total social planning. In their attitude, which I got into some trouble with scientists for publicly attacking, I noted a lack of interest in freedom, except for themselves. Along with that ran a total failure to see the dangers inherent in absolute power. Not, of course, that such attitudes are confined to scientists. I was made the more conscious of what was going on by the fact that I was at that time lecturing on Plato's *Republic*, in which the ideal of total planning was first expounded. But whereas most humanists had come to be very conscious of Lord Acton's famous dictum: "All power tends to corrupt and absolute power corrupts absolutely"— conscious of it to such a degree that they commonly left out the "tends to" in "all power tends to corrupt"—and were aware, too, that writers, artists, philosophers, historians had no freedom whatever in the Soviet Union, most scientists seemed to be indifferent either to Acton's warning or to Soviet realities.

It is by no means surprising that the first great promulgator of what we now think of as the ideal of a science-based technology, as indeed of an experimental, and so far technological, science, was an English Chancellor, Francis Bacon. His gnomic utterance "knowledge is power" associates the two very intimately; not only does knowledge give us power but power rests on knowledge, whether on a network of spies and informers or on knowledge of a scientifically engendered kind. Scientific power was at first exercised over nature. In that form, it brought humanity many benefits. Toward nature itself it was, however, ruthless. Those parts of nature which were valuable human possessions—its livestock, its farms, its gardens—it no doubt helped human beings to tend more carefully. But whenever nature stood in man's way, science provided the means for human power to be exercised in a manner that has often turned out to be disastrous not only to our fellow-inhabitants of the biosphere but to human beings themselves. So much, I scarcely need nowadays to emphasize.

There is another aspect, however, to this use of science-based power. What we call man's power over nature, C. S. Lewis once remarked, is in reality man's power over man. Science gives that kind of power to human beings which lets them more readily dominate other men, even when the power in question is in itself power over nature. That situation is exacerbated when the power is more directly and obviously power over human beings. The supreme example is the power to control human reproduction.

In Plato's *Republic* that power to control human reproduction assumes a central position in sustaining his ideal society. But he could not envisage it, of course, as consisting in anything more than a strict control over who was to be allowed to reproduce children and with whom, along with the immediate destruction of any imperfect children born of such permitted parents. Such doctrines, in the form of positive and negative eugenics, were quite widespread in the earlier decades of the present century. One tends to associate them with Hitler's Germany, but Hitler partly took them over from the United States. Now, of course, the power genetic engineering has placed in the hands of human beings goes far beyond the wildest dreams of earlier eugenicists. We find such scenarios as these seriously propounded as by the biologist Dr. James Bonner of the Californian Institute of Technology. Genetic material is taken from each individual at birth; the individual is then sterilized; a record is kept of his or her lifetime accomplishments; at his or her death a committee decides whether these accomplishments are worthy of procreation; if so, the genetic material is cloned to produce a new individual. Now, of course, we do not yet in fact possess the ability to do this; perhaps, but only perhaps, we never shall. But what always staggers me is the calm presumptions that are made in such scenarios.

One, of course, is that the environment counts for very little. For my part, I see no reason for believing that the cloning of a Shakespeare would produce another Shakespeare, of an Einstein another Einstein, of a Plato another Plato. Nor is there anything in their genetic background to suggest that it contained the clue to their achievement. More relevant to my present remarks is the assumption that this committee would make the kind of distinctions between who ought and who ought not to be cloned which would finally issue in the emergence in a perfect society composed of supermen. One has only to imagine such a committee set up in Stalin's Russia or Hitler's Germany to see how absurd that suggestion is.

A doctrine sometimes linked with this ideal of breeding supermen is that our present methods of procreation are animal and that to achieve our full humanity we must arise above the animal condition, substituting drugs, if need be, for what are now sensual pleasures. No wonder Aristophanes put Socrates in a basket; there could certainly be no better place for the propagators of these superman ideals, these quests for what H. G. Wells called *Men like Gods*. This is *hubris* all right; this is the attempt to rise above the human condition, and one need not suppose that it would be left to the gods to punish it; the punishment would be self-inflicted. I am a friend of science; I admire not only its achievements but its spirit. That some scientists are egoistic, mercenary, cruel, unscrupulous does not disturb me; that is what human beings are like under even the best of circumstances. What sometimes frightens me, rather, is their optimism, their tendency to underrate the potency of evil and, above all, of the likelihood that the evil will gravitate toward positions of power. And even beyond that, their failure to recognize the truth in what has come to be my favorite quotation: "The road to Hell is paved with good intentions."

# 4

# Emancipation Through Knowledge*

## Karl Popper

The philosophy of Immanuel Kant, and with its philosophy of history, is often looked upon in Germany as antiquated, and as superseded by Hegel and his followers. This may well be due to the surpassing intellectual and moral stature of Kant, Germany's greatest philosopher; for the very greatness of his achievement was a thorn in the flesh of his lesser successors, so that Fichte, and later Hegel, tried to solve this irritating problem by persuading the world that Kant had been merely one of their forerunners. But Kant was nothing of the sort. On the contrary, he was a determined opponent of the whole Romantic Movement and especially of Fichte: Kant was in fact the last great exponent of that much reviled movement, the Enlightenment. In an important essay entitled "What is Enlightenment?' (1785) Kant wrote:

> Enlightenment is the emancipation of man from a state of self-imposed tutelage. This state is due to his incapacity to use his own intelligence

*A broadcast delivered in German on the Bavarian Broadcasting Network in February 1961, in a series of broadcasts "On the Meaning of History." English text by the author. Originally published in Karl Popper, *In Search of a Better World* (London: Routledge, 1992). Reprinted with permission.

without external guidance. Such a state of tutelage I call "self-imposed" [or "culpable"] if it is due not to lack of intelligence but to lack of courage or determination to use one's own intelligence without the help of a leader. *Sapere aude!* Dare to use your own intelligence! This is the battle-cry of the Enlightenment.

This passage from Kant's essay explains what was for him the central idea of the Enlightenment. It was the idea of *self-liberation through knowledge.*

This idea of self-liberation or self-emancipation through knowledge remained for Kant a task as well as a guide throughout his life; and although he was convinced that this idea might serve as an inspiration for every man possessed of the necessary intelligence, Kant did not make the mistake of proposing that we adopt self-emancipation through knowledge, or any other mainly intellectual exercise, as the whole meaning or purpose of human life. Indeed, Kant did not need the assistance of the Romantics for criticizing pure reason, nor did he need their reminders to realize that man is not purely rational; and he knew that mere intellectual knowledge is neither the best thing in human life, nor the most sublime. He was a pluralist who believed in the variety of human experience and in the diversity of human aims; and being a pluralist, he believed in an open society—a pluralist society that would live up to his own maxim: "Dare to be free, and respect the freedom and the autonomy of others; for the dignity of man lies in his freedom, and in his respect for the people's autonomous and responsible beliefs, especially if these differ widely from his own." Yet in spite of his pluralism he saw in intellectual self-education, or self-emancipation through knowledge, a task which is indispensable from a philosophical point of view; a task demanding of every man immediate action here and now and always. For only through the growth of knowledge can the mind be liberated from its spiritual enslavement: enslavement by prejudices, idols, and avoidable errors. Thus the task of self-education, through certainly not exhausting the meaning of life, could, he thought, make a decisive contribution toward it.

The analogy between the expressions *the meaning of life* and *the meaning of history* is worthy of examination; but I shall first examine the ambiguity of the word "meaning" in the expression "the meaning of life." This expression is sometimes used in the sense of a deeper, a hidden meaning—something like the hidden meaning of an epigram, or of a poem, or of the *Chorus Mysticus* in Goethe's *Faust*. But the wisdom of some poets and perhaps also of some philosophers has taught us that the phrase "the meaning of life" can be understood in a different way; that the meaning of life may not be so much something hidden and perhaps discoverable but, rather, something with which we ourselves can endow our lives. We can bestow a meaning upon our lives through our work, through our active conduct, through our whole way of life, and through the attitude we adopt toward our friends and our fellow men and toward the world. (Of course, that we can endow our lives in this way may strike us as an important discovery.)

In this way the quest for the meaning of life turns into an ethical question—the question "What tasks can I set myself in order to make my life meaningful?" Or as Kant puts it: "What should I do?" A partial answer to this question is given in Kant's ideas of freedom and autonomy, and of a pluralism which is limited only by the idea of equality before the law and of mutual respect for the freedom of others; ideas which, like the idea of self-emancipation through knowledge, can contribute meaning to our lives.

We can understand the expression "the meaning of history" in a similar way. This, too, has been often interpreted in the sense of a secret or hidden meaning, underlying the course of world history; or perhaps of a hidden direction or evolutionary tendency which is inherent in history; or of a goal toward which the world is striving. Yet I believe that the quest for the hidden "meaning of history" is misconceived, as is the quest for the hidden meaning of life: instead of searching for a hidden meaning of history, we can make it our task to *give* it a meaning. We can try to give an aim to political history—and thereby to ourselves. Instead of looking for a deeper, a hidden meaning in political history, we can ask oursleves what could be worthy and humane aims of political history: aims both feasible and beneficial to mankind.

My *first thesis* is, therefore, that we should refuse to speak of the meaning of history in the sense of something concealed in it, or of a moral lesson hidden in the divine tragedy of history, or in the sense of some evolutionary tendencies or laws of history, or of some other meaning which might perhaps be discovered by some great historian or philosopher or religious leader.

Thus my first thesis is negative. I contend that there is no hidden meaning in history, and that those historians and philosophers who believe they have discovered one are deceiving themselves (and others).

My *second thesis,* however, is very positive. I believe that we ourselves can try to give a meaning to political history—or rather a plurality of meanings; meanings that are feasible for, and worthy of, human beings.

But I go even further than that. For my *third thesis* is that we can learn from history that the attempt to give history an ethical meaning, or to set ourselves up as modest ethical reformers, need not be vain. On the contrary, we shall never understand history if we underrate the historical power of ethical aims. No doubt they often have led to terrible results, unforeseen by those who first conceived them. Yet in some respects we have approached more closely than any previous generation to the aims and ideals of the Enlightenment, as represented by the American Revolution, or by Kant. More especially the idea of self-emancipation or self-liberation through knowledge, the idea of a pluralist or open society, and the idea of ending the frightful history of wars by the establishment of eternal peace, through perhaps still far distant ideals, have become the aim and the hope of almost all of us.

By saying that we have got nearer to these aims I am not, of course, venturing to prophesy that we shall soon, or ever, attain them. Certainly we may fail. But I think that at least the idea of peace which Erasmus of Rotterdam, Immanuel Kant, Friedrich Schiller, Bentham, the Mills and Spencer, and in Germany Berta von Suttner and Friedrich Wilhelm Förster, have fought for, is nowadays openly acknowledged by the diplomats and politicians of all civilized states as the aim of international politics. This is more than those great fighters for the idea of peace expected, and it is more than we could have expected even twenty-five years ago.

Admittedly, this great success is only a very partial one, and it has been brought about not so much by the ideas of Erasmus or of Kant as by the realization that a nuclear war would put an end to mankind.

But that does not alter the fact that peace is now generally and openly recognized as our political aim, and that our difficulties are mainly due to the failure, so far, of diplomats and politicians to find the means to its realization. I cannot discuss those difficulties here; yet a more detailed explanation and discussion of my three theses might make it possible to understand the difficulties and to see them in perspective.

My *first thesis,* the negative assertion that there is no hidden meaning in political history—no meaning that we might look for and discover—nor anything like a hidden tendency, contradicts the various *theories of progress* of the nineteenth century—for example the theories of Comte, Hegel, and Marx. But it also contradicts Oswald Spengler's twentieth-century theory of the *decline of the West* as well as the classical theories of *cycles* propounded for example by Plato, Giovanni Battista Vico, Nietzsche, and others.

I regard all these theories as wrong-headed, and even, in a way, pointless. For they answer a question that is wrongly put. Ideas such as "progress," "retrogression," "decline," etc., imply judgments of value; and thus all these theories, whether they predict historical progress or retrogression, or a cycle consisting of progress and retrogression, must necessarily refer to some scale of values. Such a scale of values can be moral, or economic, or perhaps aesthetic or artistic; and within the realm of the latter value it can refer to music or painting or architecture or literature. It may also refer to the realms of science, or of technology. Another scale of values may be based upon the statistics of our health or mortality, and another on our morality. Obviously, we can progress in one or several of these fields and, *at the same time,* retrogress and reach rock-bottom in others. (Thus we find in Germany at the time of our greatest works of Bach, 1720–50, no very outstanding works in literature or in painting.) And progress in some fields—say in the fields of economics or of education—must often be paid for with retrogression in others; just as progress in the speed, spread, and frequency of motor traffic is paid for at the expense of safety.

Now what is true of the realization of technological or economic values also holds, of course, for the realization of certain moral values and especially for the fundamental postulates of freedom and human dignity. Thus many citizens of the United States felt that the continuation of slavery in the Southern States was intolerable, and incompatible with the demands of their conscience; but they had to pay for the freeing of the slaves with a most terrible civil war, and with the destruction of a flourishing and unique civilization.

Similarly, the progress of science—itself partly a consequence of the ideal of self-emancipation through knowledge—is contributing to the lengthening and to the enrichment of our lives; yet it has led us to spend those lives under the threat of an atomic war, and it is doubtful whether it has on balance contributed to the happiness and contentment of man.

The fact that we can progress and simultaneously retrogress shows that the historical theories of progress, the theories of retrogression, the theories of cycles, and even the prophecies of doom, are all equally untenable, since they are clearly wrong in the way they pose their questions. They all are in the grip of pseudo-scientific theories (as I have tried to show elsewhere).[1] These pseudo-scientific theories of history, which I have called *historicist* theories, have a rather interesting history of their own.

Homer's theory of history—like that in Genesis—interprets historical events as the immediate expression of the erratic will of some highly capricious man-like deities. This type of theory was incompatible with the conception of God prevailing in later Judaism and Christianity. And indeed, to view political history—the history of robbery, war, plunder, pillage, and of ever-increasing means of destruction—as the direct work of God is nothing short of blasphemy. If history is the work of a merciful God, it can be so only if His will is for us inscrutable, incomprehensible, and unfathomable. This makes it impossible for us to understand the meaning of history, if we try to see in history the direct action of a merciful God. Thus a religion that tries to make the meaning of history really comprehensible to us (rather than leave it inscrutable) must try to understand it not as a direct revelation of the divine will of an omnipotent God but as a struggle between some good and some evil powers—powers that act in us and through us. That is what St. Augustine tried to do in his book *De Civitate Dei.* He was influenced not only by the Old Testament but also by Plato, who interpreted political history as the history of the fall from grace of an originally divine, perfect, harmonious, and communist city state, whose moral decline was caused by racial degeneration and its consequences: the worldly ambition and selfishness of the leading aristocracy. Another important influence on the work of St. Augustine derives from his own Manichean period: from the Persian-Manichean heresy which interpreted this world as an arena for the struggle between the good and the evil principles, personified by Ormuzd and Ahriman.

These influences led St. Augustine to describe the political history of mankind as the struggle between the good principle of the *civitas dei*, and the evil principle of the *civitas diaboli*—that is, between heaven and hell. And almost all later theories of history—possibly with the exception of some of the more naive theories of progress—can be traced back to this almost Manichean theory of St. Augustine. Most of the modern historicist theories simply translate his metaphysical and religious categories into the language of natural or social science. Thus they may merely replace God and the Devil by morally or biologically good races, or races fit to rule, and morally or biologically bad, or unfit, races; or by good classes and bad classes—proletarians and capitalists. ("We Communists believe," writes Khrushchev in about 1970, "that capitalism is a hell in which laboring people are condemned to slavery."[2]) This hardly alters the character of Augustine's theory.

The little that may be allowed to be correct in these theories is their inherent assumption that our own ideas and ideals are powers that influence our history. But it is important to realize that good and noble ideas may sometimes have a disastrous influence on history; and that, conversely, we can sometimes find an idea, a historical power, which wills the Bad and works the Good (as Bernard de Mandeville was perhaps the first to see); just as we can often find that an error leads to the discovery of truth.

So we must guard carefully against viewing our highly pluralist history as a drawing in black and white, or as a picture painted in a few contrasting colors. And we must be even more careful not to read into it historical laws that can be used for the prediction of progress, cycles, or doom, or for any similar historical prediction.

Yet unfortunately the general public expects and demands, especially since Hegel, and still more since Spengler, that a real scholar—a sage or a philosopher or a historian—should be able to play the role of an augur or soothsayer: that he should be able to predict the future. And what is even worse, this demand creates its own supply. In fact, the insistent demand has produced quite a glut of prophets. Without much exaggeration one could say that nowadays every intellectual of repute feels an irresistible obligation to become an expert in the art of historical prophecy. And the abysmal depth of his pessimism (for not to be a pessimist would be almost a breach of professional etiquette) is matched by the abysmal profundity and the general impressiveness of his oracular revelations.

I think it is high time to make an attempt to keep soothsaying where it belongs—in the fairground. I do not of course mean to say that soothsayers never predict the truth: if their predictions are sufficiently vague, the number of their true predictions will even exceed that of their false ones. All I assert is that there does not exist a scientific or historical or philosophical method which might help us to produce anything like those ambitious historical predictions for which Spengler created so great a demand.

Whether a historical prediction will come true or not is neither a matter of method, nor of wisdom or intuition: it is purely a matter of chance. These predictions are arbitrary, accidental, and unscientific. But any of them may well achieve a powerful propagandist effect. Provided a sufficient number of people believe in the decline of the West, the West will decline; even if, without that propaganda for its decline, it would have continued to flourish. Prophets, even false prophets, can move mountains; and so can ideas, even wrong ones. Fortunately there may be occasions when it is possible to fight wrong ideas with right ones.

In what follows I shall express some rather optimistic ideas; but they are, most emphatically, not to be taken as predictions of the future, for I do not know what the future holds, and I do not believe in those who believe they do. I am optimistic only about our ability to learn from the past and the present; we can learn that many things, both good and bad, have been and still are possible, and that we have no reason to give up hoping, striving, and working for a better world.

My *second thesis* was that we can give a meaning and set an aim to political history, a meaning and an aim or several meanings and aims, which are beneficent and humane.

Giving a meaning to history can be understood in two different ways: the more important and fundamental one is *proposing an aim* based on our ethical ideas. In another and less fundamental sense of the expression "giving a meaning," a Kantian philosopher, Theodor Lessing, has described the writing of history as *The Giving of Meaning to the Meaningless (Geschichte als Sinngebung des Sinnlosen)*. Theodor Lessing's thesis (with which I am inclined to agree even though it differs from mine) is this: we may read a meaning into the written, traditional books of history, even though history is meaningless in itself; for example, by asking how our ideas—say, the idea of freedom and the idea of self-emancipation through knowledge—have fared throughout history's tortuous course. If we are careful not to use the word "progress" in the sense of a "law of progress," we may even give a meaning to traditional history by asking what "progress" we have made, or what setbacks we have suffered, and especially what price we had to pay for making progress in certain directions. Part of the price we have paid is revealed by the history of our many tragic errors —errors in our aims and errors in our choice of means.

A similar idea has been beautifully expressed by H. A. L. Fisher, the great English historian who rejected historicism and with it all the alleged laws of historical evolution, yet who did not shrink from judging events in history from a critical point of view, applying to them the yardstick of ethical, economic, and political progress. Fisher wrote[3]:

> Men wiser and more learned than I have discerned in history a plot, a rhythm, a predetermined pattern. . . . I can see only one emergency following upon another as wave follows wave, only one great fact with respect to which, since it is unique, there can be no generalizations, only one safe rule for the historian: that he should recognize . . . the play of the contingent and the unforeseen.

Here Fisher states that there are no intrinsic developmental tendencies. Yet he continues as follows:

This is not a doctrine of cynicism and despair. The fact of progress is written plain and large on the pages of history; but progress is not a law of nature. The ground gained by one generation may be lost by the next.

Thus some progress—by progress Fisher means here social better-ment in the fields of freedom and justice, and also economic progress —may occur in spite of the senseless and cruel emergencies of war or power-political strife. But since there are no historical laws that might ensure the continuation of this progress, the future fate of progress— and with it our own fate—will largely depend on ourselves.

I have quoted Fisher not only because I believe that he is right, but also because I want to show how his idea that history depends, in part, on ourselves, is much more "meaningful" and "noble" than the idea that history has its inherent, and inexorable, laws—whether mechanical, dialectical, or organic; or that we are puppets in a historical puppet-show; or victims of superhuman historical powers, such as the powers of Good and Evil; or perhaps victims of the collective forces of proletarians and capitalists.

Thus in writing and reading history, or books of history, we can give a meaning to it. But now I turn to the other and more important sense of "giving a meaning to history": I mean the idea that we can set ourselves tasks; not only as individuals living personal lives, but also as citizens and, particularly, as citizens of the world, who regard the senseless tragedy of history as intolerable and see in it a challenge to do our best to make future history meaningful. The task is immensely difficult, mainly because good intentions and good faith can led us tragically astray. And because I support the ideas of the Enlightenment, of self-emancipation through knowledge, and of a critical rationalism, I feel it all the more necessary to emphasize the point that even the ideas of the Enlightenment and of rationalism have led to the most terrible consequences.

It was Robespierre's rule of terror that taught Kant, who had welcomed the French Revolution, that the most heinous crimes can be committed in the name of liberty, equality, and fraternity; crimes just as heinous as those committed in the name of Christianity during the Crusades, the various eras of witch hunting, and the Thirty Years' War. And with Kant we may learn a lesson from the terror of the French Revolution, a lesson which cannot be repeated too often: that fanaticism is always evil and incompatible with the aim of a pluralist society, and that it is our duty to oppose it in any form—even when its aims, though fanatically pursued, are in themselves ethically objectionable, and still more so when its aims coincide with our own personal aims. The dangers of fanaticism, and our duty to oppose it under all circumstances, are two of the most important lessons we can learn from history.

But is it possible to avoid fanaticism and its excesses? Does not history teach us that all attempts to be guided by ethical aims must be futile, just because those aims can play a historical role only when they are believed in and upheld fanatically? And does not the history of all religions and all revolutions show that the fanatical belief in an ethical idea will not only pervert it, but again and again transform it into its very opposite? That it will make us open all prison doors in the name of liberty, only to close them almost at once behind the new enemies of our new liberty? That it makes us proclaim the equality of all men, and also, that some of them "are more equal than others"? And is not this equality a jealous god who commands us to visit the inequity of some of the less equal fathers upon the chidren unto the third and fourth generation? Does it not make us proclaim the brotherhood of all men; and also, that we are the keepers of our brother— as if to remind us that our wish to rule over him may be fratricidal? Does not history teach us that all ethical ideas are pernicious, and the best of them often the most pernicious? Can we not learn from the French and Russian and more recently from some African revolutions that the ideas of the Enlightenment and the dreams of a better world are not merely nonsense, but criminal nonsense?

My answer to these questions is contained in my *third thesis:* we can learn from the history of Western Europe and the United States that the attempt to give to our history an ethical meaning or aim need not always be futile. That is not to say that we ever have realized, or ever will fully realize, our ethical aims. My assertion is very modest. All I can say is that an ethically inspired social criticism has been successful in some places, and that it has been able to eliminate, at least for the time being, some of the worst shortcomings of social and public life.

This then is my *third thesis*. It is optimistic in that it is a denial of all pessimistic views of history. For all theories of cyclical evolution, and of decline, are clearly refuted if it is possible that we ourselves impose successfully an ethical aim, an ethical meaning, upon history.

But there are certain very definite prerequisites for the imposition of ethical aims, for the successful betterment of social relations. Social ideals and social criticism were crowned by success only where people had learned to respect opinions that differ from their own, and to be sober and realistic in their political aims, where they had learned that the attempt to create the Kingdom of Heaven on earth may easily succeed in turning our earth into a hell for our fellow men.

The first countries to learn this lesson were Switzerland and England, where some utopian attempts to create a Kingdom of Heaven on earth soon led to disenchantment.

The English Revolution, the first of the great modern revolutions, did not bring about the Kingdom of Heaven but the execution of Charles I and the dictatorship of Cromwell. Thoroughly disenchanted, England learned its lesson: it was converted to believe in the need for a rule of law. The attempt of James II to reintroduce Roman Catholicism in England by force foundered on the rock of that attitude. Tired of religious and civil strife, England was ready to listen to the arguments for religious tolerance of John Locke and other pioneers of the Enlightenment, and to accept the principle that an enforced religion can have no value; that one may *guide* people into church, but must not try to *force* them into it against their convictions (as Pope Innocent XI expressed it).

The American Revolution managed to avoid the trap of fanaticism and intolerance.

It can hardly be accidental that Switzerland, England, and America, which all had to go through some disenchanting political experiences, are the countries which have succeeded in achieving, by democratic reforms, ethical-political aims which would have been unattainable by means of revolution, fanaticism, dictatorship, and the use of force.

At any rate we can learn not only from the history of the Engish-speaking democracies but also from the history of Switzerland and Scandinavia that we can set ourselves aims, and that we can sometimes achieve them—provided that these aims are neither too wide, nor too narrow, but conceived in a pluralist spirit—that is, that they embody respect for the freedom and convictions of all sorts of people with widely differing ideas and beliefs. This shows that it is not impossible to give meaning to our political history which is, precisely, my *third thesis*.

In my view it is the Romantic School and its criticism of the Enlightenment that were superficial, and not the Enlightenment, even though its name has become a synonym for superficiality. Kant and the Enlightenment were ridiculed as superficial and naive for taking seriously the ideals of liberty, and for believing that the idea of democracy was more than a transient historical phenomenon. And nowadays we can hear again a lot about the necessary transcience of these ideas. But instead of explaining their necessary transcience and prophesying their impending decline, it would be better to fight for their survival. For these ideas have not only shown their vitality, and their power to survive terrible attacks: they also have turned out to provide, as Kant thought they would, the necessary framework for a pluralist society; and vice versa: the pluralist society is the necessary framework for the working out of political meanings and aims; for any policy which transcends the immediate present; for any policy which reads a meaning into our past history, and tries to give our present and future history a meaning.

Enlightenment and Romanticism have one important point in common: both see the history of mankind mainly as a history of contending ideas and beliefs, as a history of ideological struggles. In this respect they agree. But it is in their attitude toward these ideas that Enlightenment and Romanticism diverge so widely. Romanticism values the power of faith as such: it values its vigor and depth, independently of the question of its *truth*. This it seems is the real reason why the Romantic School is so contemptuous of the Enlightenment. For the Enlightenment views faith and the power of faith with a measure of distrust. Although it teaches tolerance and even respect for other people's faith, its greatest value is not faith, but truth. And it teaches that there is something like absolute truth, even though it may be unknown to us; and that we can get nearer to it through correcting our errors. This, in fact, is the fundamental thesis of the philosophy of the Enlightenment; and in this lies its greatest contrast with the historical relativism of the Romantics.

But the approach to truth is not easy. There is only one way toward it, the way through error. Only through our errors can we learn; and only he will learn who is ready to appreciate and even to cherish the errors of others as stepping stones toward truth, and who searches for his own errors: who tries to find them, since only when he has become aware of them can he free himself from them.

The idea of our self-emancipation through knowledge is therefore not the same as the idea of our mastery over nature. The former is, rather, the idea of a spiritual self-liberation from error, from superstition and from false idols. It is the idea of one's own spiritual self-emancipation and growth, through one's own criticism of one's own ideas—though the help of others will always be needed.

Thus we see that the Enlightenment does not reject fanaticism and fanatical forms of belief for purely utilitarian reasons, or because it has found that better things can be achieved in politics and in practical affairs by a more sober attitude. Its rejection of fanatical belief is, rather, the natural corollary of the idea that we should search for truth by criticizing our errors. This self-criticism and this self-emancipation are possible only in a pluralist society, that is, in an open society which tolerates our errors as well as the errors of others.

The idea of self-emancipation through knowledge, which was the basic idea of the Enlightenment, is in itself a powerful enemy of fanaticism; for it makes us try hard to detach ourselves or even to dissociate ourselves from our own ideas (in order to look at them critically) instead of identifying ourselves with them. And the recognition of the sometimes overwhelming historical power of ideas should teach us how important it is to free ourselves from the overpowering influence of false or wrong ideas. In the interests of the quest for truth and of our liberation from errors we have to train ourselves to view our own favorite ideas just as critically as those we oppose.

This is not a concession to relativism. In fact, the very idea of error presupposes the idea of truth. Admitting that the other man may be right and that I may be wrong obviously does not and cannot mean that each man's personal point of view is equally true or equally tenable and that, as the relativists say, everybody is right within his own frame of reference, though he may be wrong within that of somebody else. In the western democracies many of us have learned that at times we are wrong and our opponents are right; but too many who have digested this important truth have slipped into relativism. In our great historical task of creating a free pluralist society, and with it a social framework for the growth of knowledge and for self-emancipation through knowledge, nothing is more vital than to be able to view our own ideas critically; without however becoming relativists or skeptics, and without losing the courage and the determination to fight for our convictions, even though we realize that these convictions should always be open to correction, and that only through correcting them may we free ourselves from error, thus making it possible for us to grow in knowledge.

## Notes

1. See *The Open Society and Its Enemies,* 9th reprint of the 5th edition, vol. I: 1991; vol. II: 1992 (London: Routledge); also *The Poverty of Historicism,* 14th impression (London: Routledge, 1991).

2. See *Khrushchev Remembers, with an Introduction, Commentary and Notes by Edward Crankshaw,* translated and edited by Strobe Talbott (London: Little, Brown and Company, 1971), especially pp. 521–22.

3. H. A. L. Fisher, *History of Europe* (1936), vol. I, p. vii.

# 5

# The Case for a
# New American Pragmatism*

## Thelma Z. Lavine

American philosophers are in a period of sobriety, of reflection upon the present situation of philosophy, in America, in the West, in the changing East. We are in the process of gaining some intellectual distance from the destructive bombshell dropped upon American philosophy by Viennese logical positivism before World War II and some intellectual distance also from the triumphal sweep of analytic philosophy across the universities of America after World War II. We are achieving as well some degree of thoughtful perspective upon the counter-tradition of phenomenology, hermeneutics, critical theory, structuralism, and deconstruction, which came to this country with the refugees from Nazism and began to be known here in the 1960s.

Our reflective distance from both these traditions is in large part a function of our growing perception of their vulnerabilities and their deficiencies, yet also of our acknowledgment of their perseverance. The recent bitter debates between contemptuous analysts and defensive pluralists in the American Philosophical Association (APA), which were

---

*Originally published in *Free Inquiry* (Summer 1991): 45–48. Reprinted with permission.

in many respects a mirror-image of the intellectual conflict between so-called Anglo-American philosophy and Continental philosophy, have become less rancorous in response to the persistent, ongoing self-criticism and mutual criticism of those traditions. In this partial clearing in which the situation in American philosophy is being reassessed, there has occurred a revival of interest in classical American pragmatism and an attempt to rescue its philosophic substance from its long submergence beneath the avalanche of analytic philosophy.

A cogent example of the current attempt to reassess philosophy in America, together with an attempt to revive interest in American pragmatism, appears in the presidential address of Richard Bernstein to the Eastern Division of the APA at its annual convention in 1988 in Washington, D.C. Richard Bernstein is pre-eminent on both counts: He is a foremost interpreter of current Continental philosophic issues as they relate to Anglo-American concerns, and is a distinguished contributor to the critical literature of American pragmatism. Bernstein undertakes to connect the two issues, American pragmatism and the present situation in American philosophy:

> I want to draw upon this [pragmatic] tradition because it enables us to gain a critical perspective on our present situation in philosophy. . . . Indeed, the pragmatic thinkers were ahead of their times. . . . If we pay close attention to the characteristic themes and challenges of the 'postmodern' discourses, we will see how they were anticipated by the pragmatists.

Bernstein proceeds to list "five interrelated substantive themes that enable us to characterize the pragmatic *ethos*": anti-foundationalism, fallibilism, the social character of the self and of community, contingency, and pluralism. It is difficult to avoid seeing that the best unqualified compliment that Bernstein can pay the pragmatists is that they anticipated developments in analytic philosophy. With the exception of the social self and critical community theme, the list is a commendation of pragmatists for being so far "ahead of their times" as to match very closely some of the important contributions of the analysts. But as Peggy Lee's poignant song of disillusionment asks, "Is that all there is?" No clearer testimony to the continued tenacious vitality of analytic philosophy is needed than that Bernstein, the author of *Beyond Objectivism and Relativism* and the empathic interpreter of phenomenologist Gadamer, should present as paradigmatic Richard Rorty's narrative of his graduate school initiation (shared by Bernstein) into analytic philosophy; or that Bernstein should use an analytic "we" in relation to "outsiders," or make reference to emigré philosophers who reshaped philosophy with positivism, but not to emigré philosophers who reshaped philosophy with the opposing tradition of phenomenology; or claim that "philosophy has been de-centered, there is no single paradigm . . . that dominates philosophy," while still giving analytic coloration to the themes he imputes to pragmatism and to the entire address.

From a somewhat similar standpoint A. J. Ayer's *Philosophy in the Twentieth Century* saw this century through the lens of the progression from logical positivism to analytic philosophy. One would never know from Ayer's book that there are, in the stormy history of twentieth-century philosophy, two great lines of modern philosophy: the Anglo-American line from Hume to Russell, logical positivism, and analytic philosophy and the Continental line from Kant to Hegel, Nietzsche, Heidegger, and Gadamer. Although Bernstein, unlike Ayer, is responsive to both traditions, nevertheless he looks at philosophy in America through an analytic lens and sees pragmatism as giving way to logical positivism and analytic philosophy, leaving a legacy of anticipations of the analysts, but bequeathing also a non-analytic call to community and mutual respect, which Bernstein hopes will heal the wound inflicted by the earlier analytic arrogance. (Bernstein does not refer to the near-mortal wounding of pragmatism.)

What is at issue here is Bernstein's (characteristically analytic) failure to identify analytic philosophy as belonging to one of the two great philosophic traditions of the twentieth century, and as deploying its claims against a counter-tradition, both situated in the wider frame of modern culture. But the formalist methods of logical positivism and the informal methods of ordinary language philosophy, both of which Bernstein cites as shaping the character of analytic philosophy, are clearly recognizable as falling within a tradition, the tradition of the Enlightenment, which arose in the seventeenth century as the first phase of the complex, pluralistic cognitive and cultural framework of Modernity that stretches into our time.

Enlightenment Modernity may be encapsulated as beginning with the scientific breakthrough of Newton, unifying the laws of terrestrial and celestial mechanics, and the political breakthrough of Locke, grounding politics upon self-evident natural rights and equality, and upon representative democracy. Both breakthroughs are founded on reason and share the claim to offer truths that are universal, absolute, realistic, and objective. Enlightenment Modernity claimed the primacy of reason in all significant domains, with substantive reason yielding true intuitions concerning human nature and society, and scientific instrumental reason yielding scientific laws of nature and technology; together, they yield a natural law of rational progress.

But by the end of the eighteenth century only instrumental reason survived skeptical challenges to the intuitions of Enlightenment Modernity. A new phase, Romantic Modernity, arose in cultural protest against the disenchanted, despiritualized, increasingly mechanized world of Enlightenment, waging the Wars of Liberation from Napoleon, the symbol of Enlightenment domination. Romantic Modernity arose as a cognitive framework linked to Enlightenment Modernity as its antithesis. The intuitions of Romanticism crystallized into a counter-framework of opposing conceptualizations: in opposition to Enlightenment primacy of reason, the primacy of spirit; in opposition to the scientific focus on fact and externality, the inward path of subjectivity; in opposition to scientific reason in its pursuit of objective and valid knowledge, the truths of history, culture, the arts, and the dialectic of personal and collective will; in opposition to the natural rights and political autonomy of the Enlightenment individual, a politics of collectivism of the left or right; in opposition to Enlightenment-style rational liberation from falsity, Romantic liberation from the hegemony of the Enlightenment mentality in its abstract, ahistorical, dehumanized universalism, absolutism, objectivism, and realism, and from the resulting bureaucratized world of government, industry, politics, and education. In Romantic Modernity we discover the counter-tradition to the Enlightenment.

Modernity is, then, cognitively pluralistic. The structure of Modernity may be seen to be a pluralistic framework, a framework that exists in the form of counter-frameworks which are constitutive of it. Modernity is the conflict and confluence of two diametrically opposed cognitive styles, each subverting, demystifying, and delegitimatizing the other's conception of human nature, truth, morality, and politics and the appropriate methodology for knowing them. Thus the heritage of Modernity is the mutual destructiveness of its component mentalities. Here is what has been identified as the "great divide" in philosophy. On the one side of the divide are the Enlightenment pursuers of rationally grounded objective, absolute, universal, and realistic truth and the analytic philosophers who are their twentieth-century descendants, deploying Enlightenment-style empirical, epistemological, logical, and linguistic arguments to attack the Enlightenment intuitions of objective, absolute, universal, and realistic truth and a rational foundation for knowledge.

On the other side of the divide are the counter-intuitions of the Romantic displacers of reason by subjectivity, group consciousness and its projections, personal and collective will, as these are constitutive of history and culture. Their twentieth-century descendants among the phenomenologists, hermeneuticists, textualists, and deconstructionists are monologic *interpretivists,* asserting the interpretive historical conceptual structures and the social webs of meaning that mediate all areas of everyday life, science, and philosophy—thus undermining Enlightenment views old and new, the old Enlightenment ahistorical, unmediated rational foundations for knowledge and the new ahistorical unmediated analytic empiricism of forms of life and their language games, ordinary language philosophy, and speech-acts.

I have attempted to situate the great divide in current philosophy historically, cognitively, and culturally as the conflict of Enlightenment and Romantic frameworks within Modernity as they remain operative at the end of the twentieth century. The divide has permeated the intellectual and scientific culture from various perspectives: the analytic and the Continental traditions; positivism as opposed by the interpretive turn; Foucault's rejection of the "blackmail of the Enlightenment"; within the methodology of the social sciences as Dallmyer's "two seedlings of Modernity, science and interpretive understanding"; the theme of revolution in the social sciences: against the dominance of positivism in sociology, Alfred Schutz's phenomenology; against Skinnerian behaviorism, the rise of cognitive psychology; against Freud's drive theory, the self-psychology of Kohut; against mainstream economics, the rise of Austrian subjectivist economics; and against mainstream physical sciences, Kuhn's historicizing paradigm theory of physics, and the current debates concerning scientific realism; Ricoeur's two-language view of Freud; and various two-language views of Marx; and Daniel Bell's *Cultural Contradictions of Capitalism.*

How have philosophers coped with this widely acknowledged divide? Three principal coping strategies may be discerned: the monologic purists; the mediators; the synthesizers.

*Monologic purists* are those who pursue a single philosophical style or methodology to the exclusion of elements from other styles or methodologies that would supply its philosophic deficiencies. Most notable here are the exclusionary styles of analytic philosophy and logical positivism and on the other side phenomenology, hermeneutics, and structural linguistics. From the Romantic side there now also appear instances of a monologic interpretivist methodology (Kuhn, Schutz, Geertz) which is expressed in Charles Taylor's triumphalist call for a "hermeneutic unity of science," displacing the late, unlamented positivist unity of science movement. Yet one of the more hopeful signs of our growing philosophical maturity is that the hermeneutic of suspicion is increasingly directed at the exclusionary tactics and philosophic distortions of monologic purity.

*Mediators* are those who attempt to discern in the conflict of frameworks the developing lines of change and the signs of convergence. Notable in performing this philosophic service are Richard Bernstein (with some ambivalence); Paul Ricoeur; Hilary Putnam; Richard Rorty; and, with singular comprehensiveness and analyticity, Joseph Margolis. The mediators, most of whom are analytic descendants of empiricist Enlightenment modernity, have in various ways agreed upon significant signs of convergence and of such a magnitude as to point to a revolution now taking place in philosophy. But this convergence comes at a cost. Richard Rorty (a serious analytic philosopher who converted to the Romantic gospel) stands forth from the others in holding to a developing convergence upon the demise of philosophy, at least in the extinction of its foundational and critical role. (Rorty echoes here

Derrida's remark that he wishes to see philosophy on the stage but no longer center-stage.)

The more moderate mediators agree on some or all of a set of convergences between the frameworks. The developing convergence is toward: opposition to foundationalism, essentialism, positivism, and naturalism; opposition also to all claims to an unmediated apprehension of truth, metaphysical, scientific, moral, and political; and to all forms of historical or structural totalizing, including the historical cognitive structures of Modernity presented in this article. Thus convergence is centered upon negativity, upon rejection of previously or currently held intuitions and conceptualizations.

The cost of convergence upon negations is that both sides of the divide are emptied out of the intuitions and conceptual structures that historically, culturally, morally, and politically have defined them. In apparent evidence of the correctness of the mediators' discernment of a negative convergence, Enlightenment analytic philosophy appears to have retreated to self-criticism, empiricist and logical (i.e., to Wittgenstein's deadly prescription for philosophy); and Romantic pluralist philosophies appear to have retreated to concrete particularism, to history of various ideas, to local cultural descriptivism (i.e., to Clifford Geertz's "local knowledge").

The mediator's convergence of the two frameworks in negativity is a convergence in loss. Lost to the Enlightenment frame are its great achievements of substantive and instrumental reason: the self-evidence and unalienability of universal individual rights as the rational ground of political democracy; the comprehensive engagement of philosophy with the sciences, in methods, validation, and foundations; the sense of historical progress in democratization, in the sciences, and in the human betterment resulting from scientific technologies. Lost unmistakably are Enlightenment intuitions of objective, absolute, realistic, and universal knowledge.

Lost on the Romantic side are creative, totalizing, world-historical visions of Novalis and Blake as these were culturally inherited by Hegel, Marx, and Dewey, all with moral and religious subtextual significance; lost are philosophic explorations of the dynamics of subjectivity; the structures of consciousness as mediating experience and knowledge;

cultural configurations and styles, the interplay of personal and collective will and of charisma and bureaucracy; the development of self-consciousness on the part of slaves and masters; and racial, ethnic, economic, religious, and gender-based social subgroups. All these philosophic achievements have been delegitimated.

We are left philosophically without the possibility of support for the foundation or coherence for our knowledge; or for a naturalistic or pragmatic comprehensive overview of the sciences, their methodologies, and their current projects; or for a legitimation of the classical liberalism of individualism and democracy; or for the philosophic aspects of historical change and the critique and diagnosis of our time (Ernest Gellner notes that only the philosophies of early pragmatism and Marxism discuss historical change at all, in a time of vast European historical and political upheavals); or the conception of philosophy as critique of culture (as Dewey argued) rather than as analytic self-criticism or as the local knowledge of contemporary Romantics. We are all left in the position of Rorty, who would like to support Harold Bloom's liberalism against the bitter pessimism of Foucault and would like to know what powerful words to say to the secret police when they come—but has no capability within his foundationless philosophy. We live then with what Hegel called the Unhappy Consciousness, a sense of lost and unreachable truths and values. There are however signs of hope:

1. Vulnerabilities on each side are now more openly recognized: on the Enlightenment analytic side, vulnerability not only to its well-known reduction of philosophy to logical and empiricist self-criticism, as has long been noted, but also to its ahistoricality and its failure to acknowledge the mediation of conceptual structures. These analytic vulnerabilities are now perceived as capable of being overcome and their deficiencies being supplied by the other side. On the Romantic side, the well-known vulnerability is to the charge of deficiency in methodological rigor and reality-testing; these vulnerabilities are now seen as capable of being overcome and their deficiencies being supplied by the other side.

2. A hopeful outcome of the acknowledgment of deficiencies on each side is the growing rejection of monologic approaches that remain untouched by their own vulnerabilities.

3. Another hopeful outcome of the mutual acknowledgment of deficiencies points toward withdrawing from the mediator's convergence upon negativity. In place of the zero-sum game of the mediators, once the theoretical incorporation of historicality and mediating cognitive structures is conceded, the binding of all positions, including the analytic, to historical conceptualization follows. The totalizing rejection by the mediators of all forms of foundationalism, essentialism, naturalism, and positivism now begins to give way to an increasingly informed comprehension of time frames and the moral and political dimensions of cognitive structures that are tied to them.

The third coping strategy is undertaken by the *synthesizers*. Synthesizers of the conflicting frameworks of Modernity are those who have generated an integrating social vision and have sought to construct a substantive social theory by selected elements derived from and satisfying the intuitions and concerns of both frameworks of Modernity. These include, notably, Marx, Weber, Mannheim, Freud, and Dewey. These great thinkers of the past failed, admittedly, to achieve integration according to the scrutiny of contemporary logical, epistemological, and methodological philosophical criticism. Yet their integrating contributions to social philosophy, social theory, and the social sciences have in large part defined these fields for the contemporary intellectual culture. Each is an exemplar of the creative possibilities within the cognitive structures of Modernity.

In our time Habermas has been engaged in the project of such an unstylish social theory, bringing Enlightenment and Romantic modes of thought "under one roof," as he says. His (Western Marxist) vision is of a rationally grounded social philosophy that would stand as a philosophical and moral bulwark against social pathologies such as that of the Nazi past and against intellectual distortions of the present. Yet his theory rests precariously on a naturalized transcendental foundationalism.

But in the present situation of philosophy, the synthesizers of the conflicting frameworks of Modernity have been the American pragmatists. Classical American philosophy came into being as an intellectual and moral response to a crisis brought on by the effects of Enlightenment modernization upon American life. The response of philosophy was to assimilate the newly available philosophic views of European Romantic

Modernity as an antithetical way of perceiving the problems of Modernity; and to integrate the two cultural styles, Enlightenment and Romantic Modernity, into a philosophy for an America whose national, legal, and cultural identity was in Enlightenment truth. Each of the classical American philosophers worked through the conflict and the integration in his own way. Insofar as it is possible to identify American philosophy, it is to perceive the characteristic form of American philosophy in the attempt to integrate the Enlightenment and Romantic modes of thought. What is characteristic of American pragmatism is its incorporativeness, its attempt to hold together the Enlightenment universality of human rights, science, and technology and the Romantic expressiveness of personal and group life. Each of the classical American philosophers was attempting to provide an enriching reconciliation of these cognitive structures of the modern world, which would constitute a unifying public philosophy for America.

American thought was thrown into interpretivism from the beginning, having to reconstruct religious beliefs and the components of the British Enlightenment within the new problematic situation of America. The pragmatist mode of interpretation (most explicit in Dewey) was the result: a contextual strategy of interpretation, which embraces the problematic context of changing material, historical, ideational, and social conditions, the engendering of conceptual structures which provide resolution, and the scrutiny of consequences.

What characterizes American pragmatism is not the analyst's ahistorical themes of fallibilism, contingency, and plurality, but its incorporation of Enlightenment explanation and Continental understanding. Is the first and proudest claim of pragmatism that it anticipated the anti-foundationalism of the analysts? To the contrary: Granted, there is no a priori—except for the pragmatic a priori of the frameworks of historical culture. Now that the analysts are beginning to concede the need for theoretical incorporation of historicality and mediating cognitive structures, the binding of all positions, including the analytic, to historical conceptualization follows. The structures of Modernity form such a historical conceptualization and their substantive contents constitute a pragmatic a priori. American pragmatism's proudest claim is to be the philosophic interpreter of Modernity. It is the one philosophy of the twentieth century that undertakes—nonreductively and nonmono-logically—to interpret and integrate the historical framework of Modernity in its Enlightenment and Romantic modes.

Far from having significance as an anticipation of analytic philosophy, pragmatism presents an interpretation of Modernity that situates analytic philosophy itself within the development of one of the structures of Modernity, Enlightenment Modernity. And pragmatism displaces the analytic convergence upon rejection of previously or currently held truths and values with the methodological requirement that all truths and values be subject to analysis and critique in terms of their contexts and consequences within the ongoing problematic situations of the culture.

Moreover, external to the arguments of philosophers, vast changes are taking place in the scientific and political worlds, outside of the intellectual purview of the analysts. In the scientific world, largely ignored by both sides of the divide, vast changes are taking place in theoretical physics, microbiology, and medical technology, and vast global, ecological and environmental problems now confront scientific technology. In the political world, there is the widespread eruption of political revolution for freedom and against communist totalitarianism, with geopolitical implications. These scientific and political changes, which carry global significance for scientific truth and political freedom, cannot leave contemporary philosophers untouched. The revitalization of philosophic synthesizing may be anticipated, with a newly informed scientific and historical vision, integrating the cognitive, moral, and political concerns of both evolving frameworks of Modernity. One can only imagine how Dewey would respond to the unified field theory in physics or the current European eruption of democracy against communist totalitarianism.

The task of any social philosophy may be said to be to reappropriate, interpret, and evaluate the intellectual and cultural structures of its time and place. It is time to see American pragmatism as the reappropriation, interpretation, and critique of the opposing cognitive structures of the historical culture and the long horizon of Modernity.

It is time for a new American pragmatism.

# 6

# Trotsky and the Wild Orchids*

## Richard Rorty

If there is anything to the idea that the best intellectual position is one which is attacked with equal vigor from the political right and the political left, then I am in good shape. I am often cited by conservative culture warriors as one of the relativistic, irrationalist, deconstructing, sneering, smirking intellectuals whose writings are weakening the moral fiber of the young. Neal Kozody, writing in the monthly bulletin of the Committee for the Free World, an organization known for its vigilance against symptoms of moral weakness, denounces my "cynical and nihilistic view" and says, "It is not enough for him [Rorty] that American students should be merely mindless; he would have them positively mobilized for mindlessness." Richard Neuhaus, a theologian who doubts that atheists can be good American citizens, says that the "ironist vocabulary" I advocate "can neither provide a public language for the citizens of a democracy, nor contend intellectually against the enemies of democracy, nor transmit the reasons for democracy to the next generation." My criticisms of Allan Bloom's *The Closing of the American Mind* led Harvey Mansfield—appointed by President Bush to the

*Originally published in *Common Knowledge* 1, no. 3 (Winter 1992): 140–53. Reprinted with permission.

National Council for Humanities—to say that I have "given up on America" and that I "manage to diminish even Dewey." (Mansfield recently described Dewey as a "medium-sized malefactor.") His colleague on the Council, my fellow philosopher John Searle, thinks that standards can only be restored to American higher education if people abandon the views on truth, knowledge, and objectivity that I do my best to inculcate.

Yet Sheldon Wolin, speaking from the left, sees a lot of similarity between me and Allan Bloom: both of us, he says, are intellectual snobs who care only about the leisured, cultured elite to which we belong. Neither of us has anything to say to blacks, or to other groups who have been shunted aside by American society. Wolin's view is echoed by Terry Eagleton, Britain's leading Marxist thinker. Eagleton says that "in [Rorty's] ideal society the intellectuals will be 'ironists,' practicing a suitably cavalier, laid-back attitude to their own belief, while the masses, for whom such self-ironizing might prove too subversive a weapon, will continue to salute the flag and take life seriously." *Der Spiegel* said that I "attempt to make the yuppie regression look good." Jonathan Culler, one of Derrida's chief disciples and expositors, says that my version of pragmatism "seems altogether appropriate to the age of Reagan." Richard Bernstein says that my views are "little more than an ideological *apologia* for an old-fashioned version of cold war liberalism dressed up in fashionable 'post-modern' discourse." The left's favorite word for me is "complacent," just as the right's is "irresponsible."

The left's hostility is partially explained by the fact that most people who admire Nietzsche, Heidegger, and Derrida as much as I do— most of the people who either classify themselves as "postmodernist" or (like me) find themselves thus classified willy-nilly—participate in what Jonathan Yardley has called the "America Sucks Sweepstakes." Participants in this event compete to find better, bitterer ways of describing the United States. They see our country as embodying everything that is wrong with the rich post-Enlightenment West. They see ours as what Foucault called a "disciplinary society," dominated by an odious ethos of "liberal individualism," an ethos which produces racism, sexism, consumerism, and Republican presidents. By contrast, I see America pretty much as Whitman and Dewey did, as opening a prospect to illimitable democratic vistas. I think that our country—despite its past and present atrocities and vices, and despite its continuing eagerness to elect fools and knaves to high office—is a good example of the best kind of society so far invented.

The right's hostility is largely explained by the fact that rightist thinkers don't think that it is enough just to *prefer* democratic societies. One also has to believe that they are Objectively Good, that the institutions of such societies are grounded in Rational First Principles. Especially if one teaches philosophy, as I do, one is expected to tell the young that their society is not just one of the better ones so far contrived, but one which embodies Truth and Reason. Refusal to say this sort of thing counts as "the treason of the clerks"—as an abdication of professional and moral responsibility. My own philosophical views—views that I share with Nietzsche and Dewey—forbid me to say this kind of thing. I do not have much use for notions like "objective value" and "objective truth." I think that the so-called postmodernists are right in most of their criticisms of traditional philosophical talk about "reason." So my philosophical views offend the right as much as my political preferences offend the left.

I am sometimes told, by critics from both ends of the political spectrum, that my views are so weird as to be merely frivolous. They suspect that I will say anything to get a gasp, that I am just amusing myself by contradicting everybody else. This hurts. So I have tried, in what follows, to say something about how I got into my present position—how I got into philosophy, and then found myself unable to use philosophy for the purpose I had originally had in mind. Perhaps this bit of autobiography will make clear that, even if my views about the relation of philosophy and politics are odd, they were not adopted for frivolous reasons.

When I was twelve, the most salient books on my parents' shelves were two red-bound volumes—*The Case of Leon Trotsky* and *Not Guilty*. These made up the report of the Dewey Commission of Inquiry into the Moscow Trials. I never read them with the wide-eyed fascination I brought to books like Kraft-Ebbings' *Psychopathia Sexualis,* but I thought of them in the way in which other children thought of their family's Bible: they were books that radiated redemptive truth and moral splendor. If I were a really *good* boy, I would say to myself, I should have read not only the Dewey Commission reports, but also Trotsky's *History of the Russian Revolution,* a book I started many times but never managed to finish. For in the 1940s, the Russian Revolution and its betrayal by Stalin were, for me, what the Incarnation and its betrayal by the Catholics had been to precocious little Lutherans four hundred years before.

My father had almost, but not quite, accompanied John Dewey to Mexico as P.R. man for the Commission of Inquiry which Dewey chaired. Having broken with the American Communist Party in 1932, my parents had been classified by the *Daily Worker* as "Trotskyites," and they more or less accepted the description. When Trotsky was assassinated in 1940, one of his secretaries, John Frank, hoped that the GPU would not think to look for him in the remote little village on the Delaware River where we were living. Using a pseudonym, he was our guest in Flatbrookville for some months. I was warned not to disclose his real identity, though it is doubtful that my schoolmates at Walpack Elementary would have been interested in my indiscretions.

I grew up knowing that all decent people were, if not Trotskyites, at least socialists. I also knew that Stalin had ordered not only Trotsky's assassination but also Kirov's, Ehrlich's, Alter's, and Carlo Tresca's. (Tresca, gunned down on the streets of New York, had been a family friend.) I knew that poor people would always be oppressed until capitalism was overcome. Working as an unpaid office boy during my twelfth winter, I carried drafts of press releases from the Worker's Defense League office to Gramercy Park (where my parents worked) to Norman Thomas' (the Socialist Party's candidate for president) house around the corner, and also to A. Philip Randolph's office at the Brotherhood of Pullman Car Porters on 125th Street. On the subway, I would read the documents I was carrying. They told me a lot about what factory owners did to union organizers, plantation owners to sharecroppers, and the white locomotive engineers' union to the colored firemen (whose jobs white men wanted, now that diesel engines were replacing coal-fired steam engines). So, at twelve, I knew that the point of being human was to spend one's life fighting social injustice.

But I also had private, weird, snobbish, incommunicable interests. In earlier years these had been in Tibet. I had sent the newly enthroned Dalai Lama a present, accompanied by warm congratulations to a fellow eight-year-old who had made good. A few years later, when my parents began dividing their time between the Chelsea Hotel and the mountains of northwest New Jersey, these interests switched to orchids. Some forty species of wild orchids occur in those mountains, and I eventually found seventeen of them. Wild orchids are uncommon, and rather hard to spot. I prided myself enormously on being the only person around who knew where they grew, their Latin names, and their blooming times. When in New York, I would go to the 42nd Street Public Library to reread a nineteenth-century volume on the botany of the orchids of the eastern U.S.

I was not quite sure why those orchids were so important, but I was convinced that they were. I was sure that our noble, pure, chaste, North American wild orchids were morally superior to the showy, hybridized, tropical orchids displayed in florists' shops. I was also convinced that there was a deep significance in the fact that the orchids are the latest and most complex plants to have been developed in the course of evolution. Looking back, I suspect that there was a lot of sublimated sexuality involved (orchids being a notoriously sexy sort of flower), and that my desire to learn all there was to know about orchids was linked to my desire to understand all the hard words in Kraft-Ebbing.

I was uneasily aware, however, that there was something a bit dubious about this esotericism—this interest in socially useless flowers. I had read (in the vast amount of spare time given to a clever, snotty, nerdy, only child) bits of *Marius the Epicurean* and also bits of Marxist criticisms of Pater's aestheticism. I was afraid that Trotsky (whose *Literature and Revolution* I had nibbled at) would not have approved of my interest in orchids.

At fifteen I escaped from the bullies who regularly beat me up on the playground of my high school (bullies who, I assumed, would somehow wither away once capitalism had been overcome) by going off to the so-called Hutchings College of the University of Chicago. (This was the institution immortalized by A. J. Liebling as "the biggest collection of juvenile neurotics since the Children's Crusade.") Insofar as I had any project in mind, it was to reconcile Trotsky and the orchids. I wanted to find some intellectual or aesthetic framework which would let me—in a thrilling phrase which I came across in Yeats—"hold reality and justice in a single vision." By *reality* I meant, more or less, the Wordsworthian moments in which, in the woods around Flatbrookville (and especially in the presence of certain coralroot orchids, and of the smaller yellow lady slipper), I had felt touched by something luminous, something of ineffable importance. By *justice* I meant what Norman Thomas and Trotsky both stood for, the liberation of the weak from the strong. I wanted a way to be both an intellectual and spiritual snob and a friend of humanity—a nerdy recluse and a fighter for justice. I was very confused, but reasonably sure that at Chicago I would find out how grownups managed to work the trick I had in mind.

When I got to Chicago (in 1946), I found that Hutchins, together with his friends Mortimer Adler and Richard McKeon (the villain of Pirsig's *Zen and the Art of Motorcycle Maintenance*), had enveloped much of the University of Chicago in a neo-Aristotelian mystique. The most frequent target of their sneers was John Dewey's pragmatism. That pragmatism was the philosophy of my parents' friend Sidney Hook, as well as the unofficial philosophy of most of the other New York intellectuals who had given up on dialectical materialism. But according to Hutchins and Adler, pragmatism was vulgar, "relativistic," and self-refuting. As they pointed out over and over again, Dewey had no absolutes. To say, as Dewey did, that "growth itself is the only moral end," left one without a criterion for growth, and thus with no way to refute Hitler's suggestion that Germany had "grown" under his rule. To say that truth is what works is to reduce the quest for truth to the quest for power. Only an appeal to something eternal, absolute, and good—like the God of St. Thomas or the "nature of human beings" described by Aristotle—would permit one to answer the Nazis, to justify one's choice of social democracy over fascism.

This quest for stable absolutes was common to the neo-Thomists and to Leo Strauss, the teacher who attracted the best of the Chicago students (including my classmate Allan Bloom). The Chicago faculty was dotted with awesomely learned refugees from Hitler, of whom Strauss was the most revered. All of them seemed to agree that something deeper and weightier than Dewey was needed if one was to explain why it would be better to be dead than to be a Nazi. This sounded pretty good to my fifteen-year-old ears. For moral and philosophical absolutes sounded a bit like my beloved orchids—numinous, hard to find, known to only a chosen few. Further, since Dewey was a hero to all the people among whom I had grown up, scorning Dewey was a convenient form of adolescent revolt. The only question was whether this scorn should take a religious or a philosophical form, and how it might be combined with striving for social justice.

Like many of my classmates at Chicago, I knew lots of T. S. Eliot by heart. I was attracted by Eliot's suggestions that only committed Christians (and perhaps only Anglo-Catholics) could overcome their unhealthy preoccupation with their private obsessions, and to serve their fellow humans with proper humility. But a prideful inability to believe what I was saying when I recited the General Confession gradually led me to give up on my awkward attempts to get religion. So I fell back on absolutist philosophy.

I read through Plato during my fifteenth summer, and convinced myself that Socrates was right—virtue *was* knowledge. That claim was music to my ears, for I had doubts about my own moral character and a suspicion that my only gifts were intellectual ones. Besides, Socrates *had* to be right, for only then could one hold reality and justice in a single vision. Only if he were right could one hope to be both as good as the best Christians (such as Alyosha in *The Brothers Karamazov,* whom I could not—and still cannot—decide whether to envy or despise) and as learned and clever as Strauss and his students. So I decided to major in philosophy. I figured that if I became a philosopher I might get to the top of Plato's "divided line"—the place "beyond hypotheses" where the full sunshine of Truth irradiates the purified soul of the wise and good: an Elysian field dotted with immaterial orchids. It seemed obvious to me that getting to such a place was what everybody with any brains really wanted. It also seemed clear that Platonism had all the advantages of religion, without requiring the humility which Christianity demanded, and of which I was apparently incapable.

*    *    *

For all these reasons, I wanted very much to be some kind of Platonist, and from fifteen to twenty I did my best. But it didn't pan out. I could never figure out whether the Platonic philosopher was aiming at the ability to offer irrefutable argument—argument which rendered him able to convince anyone he encountered of what he believed (the sort of thing Ivan Karamazov was good at)—or instead was aiming at a sort of incommunicable, private bliss (the sort of thing his brother Alyosha seemed to possess). The first goal is to achieve argumentative power over others—e.g., to become able to convince bullies that they should not beat one up, or to convince rich capitalists that they must cede their power to a cooperative, egalitarian commonwealth. The second goal is to enter a state in which all your own doubts are stilled, but in which you no longer wish to argue. Both goals seemed desirable, but I could not see how they could be fitted together.

At the same time as I was worrying about this tension within Platonism—and within any form of what Dewey had called "the quest for certainty"—I was also worrying about the familiar problem of how one could possibly get a noncircular justification of any debatable stand on any important issue. The more philosophers I read, the clearer it seemed that each of them could carry their views back to first principles which were incompatible with the first principles of their opponents, and that none of them ever got to that fabled place "beyond hypotheses." There seemed to be nothing like a neutral standpoint from which these alternative first principles could be evaluated. But if there were no such standpoint, then the whole idea of "rational certainty," and the whole Socratic-Platonic idea of replacing passion by reason, seemed not to make much sense.

Eventually I got over the worry about circular argumentation by deciding that the test of philosophical truth was overall coherence, rather than deducibility from unquestioned first principles. But this didn't help much. For coherence is a matter of avoiding contradictions, and St. Thomas' advice "When you meet a contradiction, make a distinction," makes that pretty easy. As far as I could see, philosophical talent was largely a matter of proliferating as many distinctions as were needed to wriggle out of a dialectical corner. More generally, it was a matter, when trapped in such a corner, of redescribing the nearby intellectual terrain in such a way that the terms used by one's opponent would seem irrelevant, or question-begging, or jejune. I turned out to have a flair for such redescription. But I became less and less certain that developing this skill was going to make me either wise or virtuous.

Since that initial disillusion (which climaxed about the time I left Chicago to get a Ph.D. in philosophy at Yale), I have spent forty years looking for a coherent and convincing way of formulating my worries about what, if anything, philosophy is good for. My starting point was the discovery of Hegel's *Phenomenology of Spirit,* a book which I read as saying: granted that philosophy is just a matter of out-redescribing the last philosopher, the cunning of reason can make use even of this sort of competition. It can use it to weave the conceptual fabric of a freer, better, more just society. If philosophy can be, at best, only what Hegel called "its time held in thought," still, that might be enough. For by thus holding one's time, one might do what Marx wanted done— change the world. So even if there were no such thing as "understanding the world" in the Platonic sense—an understanding from a position outside of time and history—perhaps there was still a social use for my talents, and for the study of philosophy.

For quite a while after I read Hegel, I thought that the two greatest achievements of the species to which I belonged were *The Phenomenology of Spirit* and *Remembrance of Things Past* (the book which took the place of the wild orchids once I left Flatbrookville for Chicago). Proust's ability to weave intellectual and social snobbery together with the hawthorns around Combray, his grandmother's selfless love, Odette's orchidaceous embraces of Swann and Jupien's of Charlus, and with everything else he encountered—to give each of these its due without feeling the need to bundle them together with the help of a religious faith or a philosophical theory—seemed to me as astonishing as Hegel's ability to throw himself successively into empiricism, Greek tragedy, Stoicism, Christianity, and Newtonian physics, and to emerge from each, ready and eager for something completely different. It was the cheerful commitment to irreducible temporality which Hegel and Proust shared —the specifically anti-Platonic element in their work—that seemed so wonderful. They both seemed able to weave everything they encountered into a narrative without asking that that narrative have a moral, and without asking how that narrative would appear under the aspect of eternity.

About twenty years or so after I decided that the young Hegel's willingness to stop trying for eternity, and just be the child of his time, was the appropriate response to disillusionment with Plato, I found myself being led back to Dewey. Dewey now seemed to me a philosopher who had learned all that Hegel had to teach about how to eschew certainty and eternity, while immunizing himself against pantheism by taking Darwin seriously. This rediscovery of Dewey coincided with my first encounter with Derrida (which I owe to Jonathan Arac, my colleague at Princeton). Derrida led me back to Heidegger, and I was struck by the resemblances between Dewey's, Wittgenstein's, and Heidegger's criticisms of Cartesianism. Suddenly things began to come together. I thought I saw a way to blend a criticism of the Cartesian tradition with the quasi-Hegelian historicism of Michel Foucault, Ian Hacking, and Alasdair MacIntyre. I thought that I could fit all these into a quasi-Heideggerian story about the tensions within Platonism.

The result of this small epiphany was a book called *Philosophy and the Mirror of Nature*. Though disliked by most of my fellow philosophy professors, this book had enough success among nonphilosophers to give me a self-confidence I had previously lacked. But *Philosophy and the Mirror of Nature* did not do much for my adolescent ambitions. The topics it treated—the mind-body problem, controversies in the philosophy of language about truth and meaning, Kuhnian philosophy of science—were pretty remote from both Trotsky and the orchids. I had gotten back on good terms with Dewey; I had articulated my historicist anti-Platonism; I had finally figured out what I thought about the direction and value of current movements in analytic philosophy. I had sorted out most of the philosophers whom I had read. But I had not spoken to any of the questions which got me started reading philosophers in the first place. I was no closer to the single vision which, thirty years back, I had gone to college to get.

As I tried to figure out what had gone wrong, I gradually decided that the whole idea of holding reality and justice in a single vision had been a mistake—that a pursuit of such a vision had been precisely what led Plato astray. More specifically, I decided that only religion—only a nonargumentative faith in a surrogate parent who, unlike any real parent, embodied love, power, and justice in equal measure—could do the trick Plato wanted done. Since I couldn't imagine becoming religious, and indeed had gotten more and more raucously secularist, I decided that the hope of getting a single vision by becoming a philosopher had been a self-deceptive atheist's way out. So I decided to write a book about what intellectual life might be like if one could manage to give up the Platonic attempt to hold reality and justice in a single vision.

That book—*Contingency, Irony and Solidarity*—argues that there is no need to weave one's personal equivalent of Trotsky and one's personal equivalent of my wild orchids together. Rather, one should try to abjure the temptation to tie in one's moral responsibilities to other people with one's relation to whatever idiosyncratic things or persons one loves with all one's heart and soul and mind (or, if you like, the things or persons one is obsessed with). The two will, for some people, coincide—as they do in those lucky Christians for whom the love of God and of other human beings are inseparable, or revolutionaries who are moved by nothing save the thought of social justice. But they need not coincide, and one should not try too hard to make them do so. So, for example, Jean-Paul Sartre seemed to me right when he denounced Kant's self-deceptive quest for certainty, but wrong when he denounced Proust as a useless bourgeois wimp, a man whose life and writings were equally irrelevant to the only thing that really mattered, the struggle to overthrow capitalism.

Proust's life and work were, in fact, irrelevant to that struggle. But that is a silly reason to despise Proust. It is as wrongheaded as Savanarola's contempt for the works of art he called "vanities." Single-mindedness of this Sartrean or Savanarolan sort is the quest for purity of heart—the attempt to will one thing—gone rancid. It is the attempt to see yourself as an incarnation of something larger than yourself (the Movement, Reason, the Good, the Holy) rather than accepting your finitude. The latter means, among other things, accepting that what matters most to you may well be something that may never matter much to most people. Your equivalent of my orchids may always seem merely weird, merely idiosyncratic, to practically everybody else. But that is no reason to be ashamed of, or downgrade, or try to slough off, your Wordsworthian moments, your lover, your family, your pet, your favorite lines of verse, or your quaint religious faith. There is nothing sacred about universality which makes the shared automatically better than the unshared. There is no automatic privilege of what you can get everybody to agree to (the universal) over what you cannot (the idiosyncratic).

This means that the fact that you have obligations to other people (not to bully them, to join them in overthrowing tyrants, to feed them when they are hungry) does not entail that what you share with other people is more important than anything else. What you share with them, when you are aware of such moral obligations, is not, I argued in *Contingency,* "rationality" or "human nature" or "the fatherhood of God" or "a knowledge of the Moral Law," or anything other than ability to sympathize with the pain of others. There is no particular reason to expect that your sensitivity to that pain, and your idiosyncratic lover, are going to fit within one big overall account of how everything hangs together. There is, in short, not much reason to hope for the sort of single vision that I went to college hoping to get.

So much for how I came to the views I currently hold. As I said earlier, most people find these views repellent. My *Contingency* book got a couple of good reviews, but these were vastly outnumbered by reviews which said that the book was frivolous, confused, and irresponsible. The gist of the criticisms I get from both left and right is pretty much the same as the gist of the criticisms aimed at Dewey by the Thomists, the Straussians, and the Marxists, back in the thirties and forties. Dewey thought, as I now do, that there was nothing bigger, more permanent, and more reliable, behind our sense of moral obligation to those in pain than a certain contingent historical phenomenon—the gradual spread of the sense that the pain of others matters, regardless of whether they are of the same family, tribe, color, religion, nation, or intelligence as oneself. This idea, Dewey thought, cannot be shown to be true by science, or religion, or philosophy—at least if "shown to be true" means "capable of being made evident to anyone, regardless of background." It can only be made evident to people whom it is not

too late to acculturate into our own particular, late-blooming, historically contingent form of life.

This Deweyan claim entails a picture of human beings as children
of their time and place, without any significant metaphysical or biological limitations on their plasticity. It means that a sense of moral obligation is a matter of conditioning rather than of insight. It also entails
that the notion of insight (in any area, physics as well as ethics) as a
glimpse of what is *there,* apart from any human needs and desires, cannot
be made coherent. As William James put it, "The trail of the human
serpent is over all." More specifically, our conscience and our aesthetic
tastes are, equally, products of the cultural environment in which we
grew up. We decent, liberal humanitarian types (representatives of the
moral community to which both my reviewers and I belong) are just
luckier, not more insightful, than the bullies with whom we struggle.

This view is often referred to dismissively as "cultural relativism."
But it is not relativistic, if that means saying that every moral view
is as good as every other. *Our* moral view is, I firmly believe, much
better than any competing view, even though there are a lot of people
whom you will never be able to convert to it. It is one thing to say,
falsely, that there is nothing to choose between us and the Nazis. It
is another thing to say, correctly, that there is no neutral, common
ground to which an experienced Nazi philosopher and I can repair in
order to argue out our differences. That Nazi and I will always strike
one another as begging all the crucial questions, arguing in circles.

Socrates and Plato suggested that if we tried hard enough we should
find beliefs which *everybody* found intuitively plausible, and that among
these would be moral beliefs whose implications, when clearly realized,
would make us virtuous as well as knowledegeable. To thinkers like
Allan Bloom (on the Straussian side) and Terry Eagleton (on the Marxist
side), there just *must* be such beliefs—unwobbling pivots that determine the answer to the question "Which moral or political alternative
is *objectively* valid?" For Deweyan pragmatists like me, history and
anthropology are enough to show that there are no unwobbling pivots,
and that seeking objectivity is just a matter of getting as much intersubjective agreement as you can manage.

Nothing much has changed in philosophical debates about whether objectivity is more than intersubjectivity since the time I went to college—or, for that matter, since the time Hegel went to seminary. Nowadays we philosophers talk about "moral language" instead of "moral experience," and about "contextualist theories of reference" rather than about "the relation between subject and object." But this is just froth on the surface. My reasons for turning away from the anti-Deweyan views I imbibed at Chicago are pretty much the same reasons Dewey had for turning away from evangelical Christianity and from the neo-Hegelian pantheism which he embraced in his twenties. They are also pretty much the reason which led Hegel to turn away from Kant, and to decide that both God and the Moral Law had to be temporalized and historicized to be believable. I do not think that I have more insight into the debates about our need for "absolutes" than I had when I was twenty, despite all the books I have read and arguments I have had in the intervening forty years. All those years of reading and arguing did was to let me spell out my disillusionment with Plato—my conviction that philosophy was no help in dealing with Nazis and other bullies—in more detail, and to a variety of different audiences.

At the moment there are two cultural wars being waged in the United States. The first is the one described in detail by my colleague James Davison Hunter in his comprehensive and informative *Culture Wars: The Struggle to Define America*. This war—between the people Hunter calls "progressivists" and those he calls "orthodox"—is important. It will decide whether our country continues along the trajectory defined by the Bill of Rights, the Reconstruction Amendments, the building of the land-grant colleges, female suffrage, the New Deal, *Brown* vs. *Board of Education,* the building of the community colleges, Lyndon Johnson's civil rights legislation, the feminist movement, and the gay rights movement. Continuing along this trajectory would mean that America might continue to set an example of increasing tolerance and increasing equality. But it may be that this trajectory could be continued only while Americans' average real income continued to rise. So 1973 may have been the beginning of the end: the end both of rising economic expectations and of the political consensus that emerged from the New Deal. The future of American politics may be just a series of increasingly blatant and increasingly successful variations on the Willie Horton spots. Sinclair Lewis's *It Can't Happen Here* may become an increasingly plausible scenario. Unlike Hunter, I feel no need to be judicious and balanced in my attitude toward the two sides in this first sort of culture war. I see the "orthodox" (the people who think that hounding gays out of the military promotes traditional family values) as the same honest, decent, blinkered, disastrous people who voted for Hitler in 1933. I see the "progessivists" as defining the only America I care about.

The second cultural war is being waged in magazines like *Critical Inquiry* and *Salmagundi*, magazines with high subscription rates and low circulations. It is between those who see modern liberal society as fatally flawed (the people handily lumped together as "postmodernists") and typical left-wing Democrat professors like myself, people who see ours as a society in which technology and democratic institutions can, with luck, collaborate to increase equality and decrease suffering. This war is not very important. Despite the conservative columnists who pretend to view with alarm a vast conspiracy (encompassing both the postmodernists and the pragmatists) to politicize the humanities and corrupt the youth, this war is just a tiny little dispute within what Hunter calls the "progressivist" ranks.

People on the postmodernist side of this dispute tend to share Noam Chomsky's view of the United States as run by a corrupt elite which aims at enriching itself by immiserating the Third World. From that perspective, our country is not so much in danger of slipping into fascism as it is a country which has always been quasi-fascist. These people typically think that nothing will change unless we get rid of "humanism," "liberal individualism," and "technologism." People like me see nothing wrong with any of these isms, nor with the political and moral heritage of the Enlightenment—with the least common denominator of Mill and Marx, Trotsky and Whitman, William James and Václav Havel. Typically, we Deweyans are sentimentally patriotic about America—willing to grant that it could slide into fascism at any time, but proud of its past and guardedly hopeful about its future.

Most people on my side of this second, tiny, up-market, cultural war have, in the light of the history of nationalized enterprises and central planning in Central and Eastern Europe, given up on socialism. We are willing to grant that welfare-state capitalism is the best we can hope for. Most of us who were brought up Trotskyite now feel forced to admit that Lenin and Trotsky did more harm than good, and that Kerensky has gotten a bum rap for the past seventy years. But we see ourselves as still faithful to everything that was good in the socialist movement. Those on the other side, however, still insist that nothing will change unless there is some sort of total revolution. Postmodernists who consider themselves post-Marxist still want to preserve the sort of purity of heart which Lenin feared he might lose if he listened to too much Beethoven.

I am distrusted by both the "orthodox" side in the important war and the "postmodern" side in the unimportant one, because I think that the "postmoderns" are philosophically right though politically silly, and that the "orthodox" are philosophically wrong as well as politically dangerous. Unlike both the orthodox and the postmoderns, I do not think that you can tell much about the worth of a philosopher's views on topics such as truth, objectivity, and the possibility of a single vision by discovering his politics, or his irrelevance to politics. So I do not think it counts in favor of Dewey's pragmatic view of truth that he was a fervent social democrat, nor against Heidegger's criticism of Platonic notions of objectivity that he was a Nazi, nor against Derrida's view of linguistic meaning that his most influential American ally, Paul de Man, wrote a couple of anti-Semitic articles when he was young. The idea that you can evaluate a writer's philosophical views by reference to their political utility seems to me a version of the bad Platonic-Straussian idea that we cannot have justice until philosophers become kings or kings philosophers.

Both the orthodox and the postmoderns still want a tight connection between people's politics and their views on large theoretical (theological, metaphysical, epistemological, metaphilosophical) matters. Some postmodernists who initially took my enthusiasm for Derrida to mean that I must be on their political side decided, after discovering that my politics were pretty much those of Hubert Humphrey, that I must have sold out. The orthodox tend to think that people who, like the postmodernists and me, believe neither in God nor in some suitable substitute, should think that everything is permitted, that everybody can do what they like. So they tell us that we are either inconsistent or self-deceptive in putting forward our moral or political views.

I take this near unanimity among my critics to show that most people—even a lot of purportedly liberated postmodernists—still hanker for something like what I wanted when I was fifteen—a way of holding reality and justice in a single vision. More specifically, they want to unite their sense of moral and political responsibility within a grasp of the ultimate determinants of our fate. They want to see love, power, and justice as coming together deep down in the nature of things, or in the human soul, or in the structure of language, or *somewhere*. They want some sort of guarantee that their intellectual acuity, and those special ecstatic moments which that acuity sometimes affords, are of some relevance to their moral convictions. They still think that virtue and knowledge are somehow linked—that being right about philosophical matters is important for right action. I think this is important only occasionally and incidentally.

I do not, however, want to argue that philosophy is socially useless. Had there been no Plato, the Christians would have had a harder time selling the idea that all God really wanted from us was fraternal love. Had there been no Kant, the nineteenth century would have had a harder time reconciling Christian ethics with Darwin's story about the descent of man. Had there been no Darwin, it would have been harder for Whitman and Dewey to detach the Americans from their belief that they were God's chosen people, to get them to start standing on their own feet. Had there been no Dewey and no Sidney Hook, American intellectual leftists of the thirties would have been as buffaloed by the Marxists as were their counterparts in France and in Latin America. Ideas do, indeed, have consequences.

But the fact that ideas have consequences does not mean that we philosophers, we specialists in ideas, are in a key position. We are not here to provide principles or foundations or deep theoretical diagnoses, or a synoptic vision. When I am asked (as, alas, I often am) what I take contemporary philosophy's "mission" or "task" to be, I get tongue-tied. The best I can do is to stammer that we philosophy professors are people who have a certain familiarity with a certain intellectual tradition, as chemists have a certain familiarity with what happens when you mix various substances together. We can offer some advice about what will happen when you try to combine or to separate certain ideas, on the basis of our knowledge of the results of past experiments. By doing so, we may be able to help you hold your time in thought. But we are not the people to come to if you want confirmation that the things you love with all your heart are central to the structure of the universe, or that your sense of moral responsibility is "rational and objective" rather than "just" a result of how you were brought up.

There are still, as C. S. Peirce put it, "philosophical slop-shops on every corner" which *will* provide such confirmation. But there is a price. To pay the price you have to turn your back on intellectual history and on what Milan Kundera calls "the fascinating imaginative realm where no one owns the truth and everyone has the right to be understood . . . the wisdom of the novel." You risk losing the sense of infinitude, and the tolerance, which result from realizing how very many synoptic visions there have been, and how little argument can do to help you choose among them. Despite my relatively early disillusionment with Platonism, I am very glad that I spent all those years reading philosophy books. For I learned something that still seems very important: to distrust the intellectual snobbery which originally led me to read them. If I had not read all those books, I might never have been able to stop looking for what Derrida calls "a full presence beyond the reach of play," for a luminous, self-justifying, self-sufficient synoptic vision.

By now I am pretty sure that looking for such a presence and such a vision is a bad idea. The main trouble is that you might succeed, and your success might let you imagine that you have something more to rely on than the tolerance and decency of your fellow human beings. The democratic community of Dewey's dreams is a community in which nobody imagines that. It is a community in which everybody thinks that it is human solidarity, rather than knowledge of something not merely human, that really matters. The actually existing approximations to such a fully democratic, fully secular community now seem to me the greatest achievements of our species. In comparison, even Hegel's and Proust's books seem optional, orchidaceous extras.

# 7

# In Defense of Secular Humanism*

## Adolf Grünbaum

During a period of considerable strife and moral turmoil in society, there is a perennial tendency in some quarters to offer ethical nostrums. Often we are told that the theistic creeds permit the resolution of our moral perplexities, whereas secular humanism only exacerbates them, leaving moral decay and the decline of our civilization in its wake. These claims have been turned into a political gospel in the United States. Gravely, William A. Rusher, the former publisher of the *National Review,* has blamed secular humanism for producing an amoral sort of human being in our inner cities:

> What is happening to us, and what can be done? Simply put, the secular humanists have been gnawing away at the foundations of Western civilization (God, morality, the family) for two centuries, and have finally succeeded in producing, especially in our inner cities, an almost totally amoral kind of human being—a sort of human pit bull. Our country will recover, if at all, only by discovering and recommitting itself to the great salvific truths on which our civilization was founded.[1]

---

*Originally published in *Free Inquiry* (Fall 1992): 30–39. Reprinted with permission. A follow-up to this article appears in *Free Inquiry* (Fall 1993).

In a free society, the purveyors of these religious nostrums have, of course, every right to preach to their own faithful, and, indeed, to make all others aware of their moral injunctions. Thus, the pope is entitled to condemn the use of so-called artificial birth control, as distinct from the rhythm method. Yet secular humanists claim entitlement to consider that prohibition barbaric not only sexually but also demographically, if only because it contributes to the population explosion and concomitant ecological ravages, especially in the third world countries of Latin America and Africa.

But nowadays, theistic moral advocacy is readily turned into political intimidation, designed to browbeat into conformity or silence those who share Sidney Hook's perception that "Whatever is wrong with Western culture, there are no religious remedies for it, for they have been all tried." Such coercive attempts are being made in our society by both Christians and Jews.

The centerpiece of the religious creeds that are purported to be essential to both private morality and good citizenship is theism: the belief in the existence of an omnibenevolent, omnipotent, and omniscient God to whose will the universe owes its existence at all times and who is distinct as well as independent from His creation. We learn that this theistic doctrine is normatively indispensable as a source of meaningful ethical prescriptions, although the attributes of omnipotence, omnibenevolence, and omniscience are incompatible. Thus, in the eighteenth century, Immanuel Kant argued that the realizability of morality, as construed by him, requires the God of theism and indeed human immortality as its underwriter. To boot, sometimes, we are told, without the slightest attempt to supply supporting statistics, that such religious belief is actually motivationally necessary, in point of empirical fact, to assure such adherence to moral standards as is found in society. In short, the religious nostrum is that religious belief is normatively, if not motivationally, indispensable to moral conduct and good citizenship in our society.

In just this vein, Henry Grunwald, a former editor-in-chief of *Time* and one-time U.S. ambassador to Austria, opined: "Secular humanism (a respectable term even though it became a right-wing swearword) stubbornly insisted that morality need not be based on the supernatural. But it gradually became clear that ethics without the sanction of some higher authority simply were not compelling."[2] To emphasize the alleged moral anarchy ensuing from secular humanism, Grunwald approvingly quotes Chesterton's dictum: "When men stop believing in God, they don't believe in nothing; they believe in anything."

This pejorative attitude toward atheism is even codified in the ethically derogatory secondary meaning of the term *atheist* given in the unabridged *Webster's Dictionary:* "A godless person; one who lives immorally as if disbelieving in God."

Furthermore, as reported in an article on "America's Holy War,"[3] it is now being argued that the separation of church and state in the U.S. has gone too far: "A nation's identity is informed by morality, and morality by faith,"[4] "faith" being faith in the god of the mainline theistic religions. This "accommodationist" position is epitomized by Chief Justice Rehnquist of the U.S. Supreme Court, who declared that the wall of separation between church and state is "based on bad history. . . . It should be frankly and explicitly abandoned."[5] Relatedly, many devout parents see evil as instantiated alike by "sex, drugs or secular humanism."[6]

Indeed, as *Time* tells us further, "such families also believe that faith is central to serious intellectual activity and should not be relegated to Sunday school."[7] One must wonder at once how intellectual titans like Bertrand Russell or Einstein, who rejected theism, ever managed to make their contributions! Fear of the alleged dire consequences of secular humanism may well also animate creationist opposition to the theory of biological evolution, which many creationists see as abetting secular humanism.[8]

For brevity and style, let the terms *religious* or *religion* refer to the theistic species of religion, i.e., theism. This usage is indeed the primary one given in *Webster's Dictionary*. The theistic religions are usually held to comprise Judaism, which is unequivocally monotheistic, trinitarian Christianity, and Islam. Christianity and Islam were successor religions to Judaism.

Yet the term *religion* is employed very ambiguously. For example, John Dewey's notion of religion is far wider than the doctrine of theism. Sometimes the term is meant to refer to the historical phenomenon of an institutionalized form of social communion involving participation in a set of ritualistic practices, in abstraction from any doctrines that may provide the rationale for them.

The theistic creeds feature claims about the existence of God, his nature, including his causal relations to the world, as well as ethical teachings that are held to codify the divine moral order of the world within the framework of the theological tenets. Yet, the appraisal of the complaints made by theists against secular humanism requires that we distinguish the theological from the moral components of their creeds in order to clarify the conceptual relations between them.

One vital lesson of that analysis will be that neither theism nor atheism as such permit the logical deduction of any judgments of moral value or of any ethical rules of conduct. Moral codes turn out to be logically extraneous to each of these competing philosophical theories. And if such a code is to be integrated with either of them in a wider system, the ethical component must be imported from elsewhere.

In the case of theism, it will emerge that neither the attribution of omnibenevolence to God nor the invocation of divine commandments enables its theology to give a cogent justification for any particular action-able moral code. Theism, no less than atheism, is itself morally sterile: Concrete ethical codes are autonomous with respect to either of them.

Just as a system of morals can be tacked onto theism, so also atheism may be embedded in a secular humanism in which concrete principles of humane rights and wrongs are supplied on other grounds. Though atheism itself is devoid of any specific moral precepts, secular humanism evidently need not be. By the same token, a suitably articulated form of secular humanism can rule out some modes of conduct while enjoining others, no less than a religious code in which concrete ethical injunctions have been adjoined to theism (e.g., "Do not covet thy neighbor's wife").

Therefore, it should hardly occasion surprise that theism is not logically necessary as one of the premises of a systematic moral code. And this failure of logical indispensability patently discredits Dostoyevsky's affirmation of it via Smerdyakov's dictum in *The Brothers Karamazov:* "If God doesn't exist, all things are permissible." Indeed, Smerdyakov's epigram boomerangs: Since atheism and theism are alike ethically barren, neither doctrine itself imposes any concrete moral prohibitions on human conduct.

The arguments supporting these contentions will undermine the rather strident attacks leveled against secular humanism in 1991 by Irving

Kristol and Richard John Neuhaus, as well as those delivered earlier by Alexander Solzhenitsyn.

Some twentieth-century theists articulated the notion of divine omnibenevolence with a view to reconciling it with what most civilized people would surely regard as great moral and natural evil. Theological apologetics—so-called theodicy—is designed to vindicate the justice and omnibenevolence of an omnipotent and omniscient God in a world of rampant evil. The pronouncements of a leading Jewish theologian and of some prominent Orthodox rabbis will show that the notion of divine omnibenevolence is shockingly permissive morally, to the point of sanctioning the justice of the Holocaust! True enough, there are other theists who would reject these fundamentalist biblical theodicies. Yet, precisely that divergence will itself be evidence for the moral hollowness of theism.

## The Problem of Evil and the Moral Permissiveness of Theism

The problem of acknowledged moral evil has perennially bedeviled those who believe in the governance of the world by a just, or even omnibenevolent God. No wonder, therefore, that the influential twentieth-century Jewish theologian Martin Buber saw the Nazi Holocaust as a particularly acute challenge to the doctrine of divine justice.

Buber felt driven to conclude that God temporarily goes into eclipse during such periods as that of the Holocaust. But just why a benevolent God would go into eclipse to accommodate the likes of Adolf Hitler, Buber could not explain. After all, going into such an eclipse would seem to be a case of morally irresponsible absenteeism on God's part. Indeed, if Buber is to be believed, and if one looks at the history of the societies that have embraced theism in one form or another, it is difficult to find any time at all when God was not at least partially in eclipse.

Buber's eclipse theodicy is not only lame but frivolous. Worse, some recent apologetics for the Holocaust from some Jewish religious quarters have been nothing less than obscene. In a 1987 article[9] Lord Immanuel Jakobovitz, the Chief Orthodox Rabbi of Britain and the Commonwealth, asserted that the Nazi Holocaust was divine punishment for the apostasy of the German Jews who founded assimilationist Reform Judaism. "This idol of individual assimilation," he wrote almost gleefully, "exploded in the very country in which it was invented, to be eventually melted down and incinerated in the crematoria of Auschwitz."

Now, when the S.S.-men who implemented the "final solution" have their reunions, they can say—on the authority of none other than the Chief Orthodox Rabbi of the United Kingdom—that they were merely the instruments of the God of Moses. Indeed, if Rabbi Jakobovitz is to be believed, the wrath of God is so indiscriminate that it prompted the Nazis to incinerate devoutly Orthodox Jews from all over Central Europe, no less than the supposedly wicked Reform Jews of Germany. Moreover, the vindictiveness of this God is such that the punishment for the doctrinal deviance of Reform Jews, even within a Mosaic theistic framework, had to be nothing short of live incineration, rather than some lesser, reversible misfortune. Far from being just, a God who indiscriminately assigns wholesale lethal punishment and allows babies to be killed in front of their mothers by S.S. guards at extermination camps is a sadistic, satanic monster deserving of cosmic loathing rather than worship and love.

Rabbi Jakobovitz is hardly alone in the view that the Holocaust was divinely sanctioned. As reported by the noted Israeli scholar Amos Funkenstein, the ultra-Orthodox Rabbi Joel Teitelbaum—who lives in Jerusalem but regards the Jewish secular state and government in Israel as sinful—sees the Holocaust as God's punishment for the Zionist founding of a Jewish state *in advance* of the promised arrival of the purported new Messiah. Evidently, Rabbi Teitelbaum also regards God's justice as morally indiscriminate. After all, many of the European Jews who perished in the crematoria were not even Zionists, let alone participating citizens of the state of Israel. And it seems to have been lost on both of the rabbis that the principle of wholesale, collective guilt and justice is invoked by Islamic terrorists who attack Israeli citizens no less than others.

Not to be outdone by rabbis Jakobovitz and Teitelbaum, the ultra-Orthodox Brooklyn Rabbi Menachem Mendel Schneerson, who is now hailed as the new Messiah by his disciples, gave his own twist to the vindication of the Holocaust. In his 1980 book *Faith and Science (Emunah v' Madah)*, this revered sage of orthodoxy opined that, in permitting the Holocaust, God cut off the gangrenous arm of the Jewish people. On this basis, this man of God concludes, the Holocaust was a good thing, because without it, the entire Jewish people would have perished. Just why that should have happened is left unclear.[10] The zealots who proclaim Schneerson to be the new Messiah suggest that the wonders he will enact are imminent. Yet, we can be sure that when these miracles fail to materialize, we will be treated to other, soothing prophecies on the model of the Barnum statements found in astrological forecasts or Chinese fortune cookies.

Sidney Hook explained why he rejected theism, including Judaism, the religion of his ancestors, in favor of atheism. In a response, the Orthodox Chicago Rabbi Yaakov Homnick[11] indicted Hook's rejection of his heritage as "a far greater tragedy than all of the physically maimed children in the world." Indeed, Rabbi Homnick goes Buber, Jakobovitz, and Teitelbaum one better in his discernment of the hand of God, which he deems patent in the Holocaust: "Yes, without a doubt, the guidance of history by G-d is perceptible even to our limited gaze. The sense of justice . . . is palpable. . . . Especially is the Holocaust a proof of G-d's justice, coming as a climax of a century in which the vast majority of Jews, after thousands of years of loyalty in exile, decided to cast off the yoke of the Torah." The rabbi's deletion of the letter *o* from the spelling of *God* is intended to convey reverence, as if the word *God* were God's true, hallowed name.

Rabbi Homnick's veritable paean to divine retribution prompted Sidney Hook[12] to reply: "All apologists, whether Christian or Jewish, for the divine inspiration of the Bible end up justifying . . . actions that in ordinary moral discourse we should regard as wicked or evil. This would be evidence enough that, in our discussions with them, we are not using terms like *good* and *bad, right* and *wrong* in the same sense." After all, Hook pointed out, these apologists "cannot really share with us a common universe of moral discourse, since they claim that every event inspired or approved by Jehovah [such as the Holocaust] is morally good."

In fact, the Bible, though called "The Good Book," features some appalling teachings ranging from genocide in Deuteronomy, to slavery and the inferior status of women in the New Testament. Besides, as Robin Lane Fox has shown,[13] the Bible contains massive historical errors and contradictions, which furnish a devastating case against strict biblical fundamentalism.

Since theological teachings lend themselves to countenancing the stated enormities, this unconscionable permissiveness provides strong reason to reject the pertinent creedal systems.

## The Moral Sterility of Theism

The moral hollowness of the theistic superstructure requires both clarification and argument. Why are theological trappings morally unavailing? It was Socrates who permitted us to realize that if a religious creed is to yield any specific moral prescriptions at all, the ethics must be extraneously imported or tacked on to theism, being put into the mouth of God by the clergy when asserting His goodness or omnibenevolence. This moral sterility of theism comes into view from the failure of divine omnibenevolence to deal with the challenge posed by a key question from Socrates in Plato's *Euthyphro:* Is the conduct approved by the gods right, because of *properties of its own,* or merely because it pleases the gods to value or command it? In the former case, divine omnibenevolence and revelation are at best ethically superfluous, and in the latter, they are unavailing.

For if God values and enjoins us to do what is desirable in its own right, then ethical rules do not depend for their validity on divine command, and they can then be independently adopted. But, on the other hand, if conduct is good merely because God decrees it, then we have the morally insoluble problem of deciding which one of the conflicting purported divine revelations of ethical commands we are to accept. Indeed, Richard Gale sees the thrust of Plato's *Euthyphro* to be the claim that "ethical propositions are not of the right categoreal sort to be made true by anyone's decision [command], even God's."[14]

The plurality of competing revelations is illustrated by those in which Jesus is the Lord and those in which he is not, as in Islam and Judaism. And how are we to resolve theologically the basic ethical disagreements existing even within the clergy of the same religious denomination, such as the debate on pacifism in times of war or the justice of capital punishment for crime? Just these conflicting moral revelations and intra-denominational disagreements spell a cardinal lesson: Even if a person is minded to defer completely to theological authority on moral matters, he or she cannot avoid deciding which one of the conflicting religious authorities is to be the ethical guide. Thus, try as they may, people cannot abdicate their own responsibility for deciding by what moral norms they are to live. In just this decision-making sense, man is inescapably the measure of all things, for better or for worse.

True enough, assuming divine omnibenevolence, it presumably follows that all divinely ordained conduct is morally right. But that is unavailing, because this much leaves us wholly in the dark as to which moral directives are binding on us, or what goals are ethically desirable. How, for example, does divine omnibenevolence tell us whether to share or abhor the Reverend Falwell's and Rabbi Kahane's claim that a nuclear Armageddon is part of God's just and loving plan for us, because only the righteous will be resurrected thereafter? In any case, the existence of states of affairs in the world that theists themselves acknowledge to be morally evil, no less than others do, does indeed impugn the purported omnibenevolence of God. And the existence of evil that is not wrought by human volition cannot be explained away by recourse to the so-called free will defense. That apologia adduces the value of human freedom to perpetuate evil deeds no less than to work good ones.

The inability of the theological superstructure to yield a moral code also crops out in Kant's invocation of God (and of personal immortality) as underpinnings of his own system of deontological ethics. His argument for such a theological foundation starts out from his moral doctrine that there is a categorical imperative to act only on principles that *everyone* could adopt consistently. But Kant avowedly offered only a formula: Alas, it does not tell us which moral directives to adopt from a set of competing ones. Thus, instead of being a source of concrete ethical injunctions, his formula provides only a necessary condition of their acceptability.

Even at that, Kant's theological underpinning of his ethics loses its force, if only because the required realizability of the highest good is hardly assured. Besides, his case for a divine underwriter founders on its dubious assumption of personal immortality. And his argument becomes baseless in the context of such rival conceptions of ethics as are offered by the teleological or self-realization schools. Indeed, even if the philosophical viability of morality were evidence for the existence of God, as claimed by Kant, the ubiquitous reality of evil in the world would be stronger evidence against theism.

It would seem that Kant's own special version of a theological foundation for ethics fails, even if one disregards the legitimacy of non-deontological systems of ethics.

Alexander Solzhenitsyn's 1978 commencement address at Harvard showed no awareness of the moral sterility of theism:

> There is a disaster which is already very much with us. I am referring to the calamity of an autonomous [despiritualized] and irreligious, humanistic consciousness. It has made man the measure of all things on Earth, imperfect man, who is never free of pride, self-interest, envy, vanity, and dozens of other defects. . . . Is it true that man is above everything? Is there no Superior Spirit above him?

*Prima facie,* this declaration may sound ingratiatingly modest. But, as it stands, it is morally hollow and theologically question-begging. Whose revelation, one must ask, is to supplant man as the measure of all things? That of the czarist Russian Orthodox church? Or the edicts of the Ayatollah Khomeini, as enforced by his mullahs? Those of the Dutch Reformed church in apartheid South Africa? Or the teachings of Pope John Paul II, who—amid starvation in Africa—is getting support from the native episcopate for the prohibition of artificial birth control? Or yet those of the Orthodox rabbinate in Israel, which prohibits autopsies, for example? And, if the latter, which of the two doctrinally *competing* chief rabbis is to be believed, the Ashkenazi or the Sephardic one? If the ethical perplexity of modern man is to be resolved by concrete moral injunctions, Solzhenitsyn's jeremiad simply replaces secular man by clergymen, who become the moral touchstone of everything by claiming revealed truth for particular ethical directives.

It appears that the moment a theology is to be used to yield ethical prescriptions, these rules of conduct are obtained by deliberations in whose outcome secular aims and thought are every bit as decisive as in the reflections of secular ethicists who deny theism. And the perplexity of moral problems is not lessened by the theological superstructure, which itself leaves us in an ethical quandary.

No wonder that Judeo-Christian theology has been invoked as a sanction for such diverse ethical doctrines as the divine right of kings; the inalienable rights of life, liberty, and the pursuit of happiness; black slavery; *"Deutschland über alles"*; the social Darwinism of Spencer; and socialism. Indeed, as Sidney Hook has pointed out in his own critique of Solzhenitsyn: "Neither Christianity nor Judaism, in principle, ever condemned slavery or feudalism. In their modern forms, they have been humanized in consequence of [the challenge from] the rise of secular humanism."

Some religious sects in India would have us abstain from the surgical excision of cancerous growths in man, and Christian Scientists in the West reach somewhat similar conclusions from rather different premises. Roman Catholics, on the other hand, endorse the medical prevention of death but condemn interference with nature in the form of birth control, a position not shared by leading Protestant and Jewish clergymen. Indeed, both Mahatma Gandhi and Hitler saw themselves as serving God. And divine Providence was a frequent feature of Hitler's speeches, illustrating anew that religion can also be the last refuge of the scoundrel. Indeed, one believer's will of God is another's will of Satan, as illustrated by the exchange between Ayatollah Khomeini and President Jimmy Carter, a born-again Christian.

Unfortunately, leading opinion-makers in the United States seem unaware not only of the moral sterility of theism, but also of the ethical abominations perpetrated by theocracies, past and present. Nor do they challenge the smug, wholly undocumented claim that the moral conduct of religiously motivated people is (statistically) superior to that of secular humanists. After all, even the members of the Mafia, for example, typically are affiliated Roman Catholics.

Solzhenitsyn's charge of moral inadequacy against an irreligious humanistic consciousness is of a piece with the point of his rhetorical questions: "Is it true that man is above everything? Is there no Superior Spirit above him?" Surely the assumption that man may well *not* be above everything hardly requires belief in the existence of God. As we know, the National Aeronautics and Space Administration has been scanning the skies for signals from extraterrestrial and indeed extrasolar humanoids, whose intelligence may indeed be superhuman.

Nor will it do for clergymen to appeal—as they often do when thus challenged by the stated damaging considerations—to the finitude of our minds or to the inscrutability of God, who is said to transcend human understanding. After all, the clergy is in no better position to transcend that finitude than anyone else. Nor, it must be emphasized, do religious apologists have greater expertise than nonbelievers for discerning the limits of human cognition. Besides, one would expect that the avowed inscrutability of God would induce great modesty in regard to fathoming his purported will and alleged ethical commands.

Those who claim a divine foundation for their favorite moral code, as against its available rivals, compensate for the ethical barrenness of theism by begging the question: They blithely claim *revealed* divine sanction for their own moral code. It was Moses, not God, who issued the Ten Commandments. The famous law code of the Babylonian King Hammurabi was purportedly received by him from the sun god Shamash during prayer, a tale similar to the legend of Moses and the revelation of the Decalogue by Yahweh on Mount Sinai. Indeed, the theological grounding of ethics is so shaky that the craving for it legitimately calls for psychological explanation as part of the psychology of fideist acceptance of theism.[15]

Irving Kristol deplores the secularization of American Jewry under the influence of secular humanism, which he tendentiously describes as springing from a "new, emergent religious impulse." As he sees it:

> Because secular humanism has, from the very beginning, incorporated the modern scientific view of the universe, it has always felt itself—and today still feels itself—"liberated" from any kind of religious perspective. But secular humanism is more than science, because it proceeds to make all kinds of inferences about the human condition and human possibilities that are not, in any authentic sense, scientific. Those inferences are metaphysical, and in the end theological.[16]

Kristol muddies the waters: Secular humanists are well aware that scientific knowledge does not suffice to warrant all parts of a moral code. But Kristol darkens counsel by designating the motivation for secular humanism as "religious," and its conception of the human estate as "theological." In so doing, he ignores that the unabridged *Webster's Dictionary* gives the following primary definition of the term *religion:* "The service and adoration of God or a god as expressed in forms of worship, in obedience to divine commands, especially as found in accepted sacred writings."

Although the term *spiritual* has a supernaturalist tinge, Kristol insists that secular humanism springs from a "new philosophical-spiritual impulse":

> What, specifically, were (and are) the teachings of this new philosophical-spiritual impulse? They can be summed up in one phrase: "Man makes himself." That is to say, the universe is bereft of transcendental meaning, it has no inherent teleology, and it is within the power of humanity to comprehend natural phenomena and to control and manipulate them so as to improve the human estate. Creativity, once a divine prerogative, becomes a distinctly human one. . . .
>
> Man's immortal soul has been a victim of progress, replaced by the temporal "self"—which he explores in such sciences as psychology and neurology, as well as in the modern novel, modern poetry, and modern psychology, all of which proceed without benefit of what, in traditional terms, was regarded as a religious dimension.[17]

First, we ought to applaud precisely what Kristol bemoaned when he said: "Creativity, once a divine prerogative, becomes a distinctly human one." The invocation of a divine creator to provide causal explanations in cosmology or biology suffers from a fundamental defect vis-à-vis scientific explanations of the effects produced by human agents or by diverse events: As we know from two thousand years of theology, the hypothesis of divine creation does not even envision, let alone specify, an appropriate intermediate causal process that would link the will of the supposed divine (causal) agency to the effects that are attributed to it. Nor, it seems, is there any prospect at all that the chronic inscrutability of the putative causal linkage will be removed by new theological developments.

In sharp contrast, the discovery that an aspirin a day keeps many a heart attack away has been quickly followed by the quest for a specification of the mode of action that mediates the prophylaxis afforded by this drug against coronary infarcts. Similarly for therapeutic benefits from placebos wrought by the mediation of endorphin release in the brain and by the secretions of interferon and of steroids. In physics, there is either an actual specification or at least a quest for the mediating causal dynamics linking presumed causes to their effects. In the case of laws of temporal coexistence or simultaneous action at a distance, there is a specification of the concomitant variations of quantified physical attributes by means of functional dependencies.[18]

Indeed, the prominent American Jesuit theologian Michael Buckley makes an important admission as to the hypothesized process of divine creation: "We really do not know how God 'pulls it off.' Catholicism has found no great scandal in this admitted ignorance."[19] But if so, the disbelief in divine creativity, which Kristol bewails, incurs no explanatory loss at all.

Kristol also deplores current disbelief in personal immortality of the soul among educated people. Yet, on examination, that tenet is so obscure that it should not be consoling to any reflective person. As Maimonides saw it, the attempt to grasp the nature of eternal bliss in the hereafter while we are alive is akin to the futile effort of a blind person to experience color visually. At any rate, the hypothesis of personal immortality collapses in the face of the vast amount of evidence for the dependence of the very existence of consciousness on adequate brain function, and, moreover, for the dependence of the integrity of our personalities on such function. Witness, for example, the effects of brain tumors, Alzheimer's disease, and various drugs, such as alcohol or mood-altering medications.[20]

But Kristol's principal thesis is that two fundamental flaws undermine the credibility of secular humanism. The first, we learn, lies in its inability to provide a moral code; the second, which is even more fundamental, is that its vision renders our lives meaningless and has become "brain dead."

As to the first, we are told:

We have, in recent years, observed two major events that represent turning points in the history of the 20th century. The first is the death of socialism, both as an ideal and a political program, a death that has been duly recorded in our consciousness. The second is the collapse of secular humanism—the religious basis of socialism—as an ideal, but not yet as an ideological program, a way of life. The emphasis is on "not yet," for as the ideal is withering away, the real will sooner or later follow suit. . . .

This loss of credibility flows from two fundamental flaws in secular humanism.

First, the philosophical rationalism of secular humanism can, at best, provide us with a statement of the necessary assumptions of a moral code, but it cannot deliver any such code itself. Moral codes evolve from the moral experience of communities, and can claim authority over behavior only to the degree that individuals are reared to look respectfully, even reverentially, on the moral traditions of their forefathers. It is the function of religion to instill such respect and reverence. Morality does not belong to a scientific mode of thought, or to a philosophical mode, or even to a theological mode, but to a practical-juridical mode. One accepts a moral code on faith—not on blind faith but on the faith that one's ancestors, over the generations, were not fools and that we have much to learn from them and their experience. Pure reason can offer a critique of moral beliefs but it cannot engender them.[21]

Elsewhere, Kristol claimed more explicitly: "Secular rationalism has been unable to produce a compelling, self-justifying moral code."[22]

These assertions call for a series of critical comments, showing that fideist theism has hardly succeeded ethically where secular rationalism has failed:

1. What is Kristol's evidence for the purported decline in adherence to secular humanism among educated people, who he tells us, had widely accepted secular humanism as an ideal? Indeed, this alleged collapse, and his prediction of its demise as an ideological program of practical action, is born of wishful thinking. Witness the well-documented massive erosion of religious belief and worship in Western Europe, which is publicly lamented by its religious leaders.

Even in the United States, where avowed religiosity is far greater than in Europe, the Roman Catholic church faces a crisis in the recruitment of young people for the priesthood. Just this scarcity of recruits has lent urgency to the plea that women be ordained as priests. The widespread disregard for the church's prohibition of artificial birth control by American Catholics is likewise well known. And the pressure to abandon the requirement of celibacy for the priesthood derives practical poignancy from the growing number of lawsuits from practicing Catholics whose children have been sexually molested by members of the clergy. On the other hand, fundamentalist Protestant evangelism is on the rise and, to the consternation of the Roman Catholic hierarchy, is making considerable inroads in certain segments of its erstwhile faithful. But this headway of fundamentalism is largely confined to the most poorly educated segment of our society. Thus, it is only cold comfort for Kristol.

2. More fundamentally, Kristol erects a straw man when he complains that the philosophical rationalism of secular humanism cannot deliver a moral code. This charge is a red herring for at least two reasons: (1) Theism as such has turned out to be morally sterile no less than atheism or "philosophical rationalism," taken by themselves; in fact, when Kristol urged that "morality does not belong to a scientific mode of thought," he himself conceded that morality *also* does not belong "even to a theological mode, but to a practical-juridical mode." (2) Secular humanism can tack on moral directives to its atheism on the basis of value judgments made by its adherents, just as, in point of actual fact, theists tack on such directives under the purported aegis of inscrutable divine revelation. Yet, unlike revelationist theists, humanists insist on the liability of their moral convictions to criticism. Kristol allowed that "Pure reason can offer a critique of moral beliefs," but his aim in saying so was not to make a partial concession; instead it was to complete the sentence by saying one-sidedly: "but it cannot engender them." Nor, as he fails to see, can theism "produce a compelling, self-justifying moral code."

Kristol draws precisely the wrong lesson from his correct observation that the erosion of belief in theism attenuated the "moral code inherited from the Judeo-Christian tradition." For, in his view, it tells against secular humanism that thereupon "we have found ourselves baffled by the Nietzschean challenge: If God is really dead, by what authority do we say [that] any particular practice is prohibited or permitted?" By now, it should be abundantly clear, however, that in answering the question as to the "authority" for concrete moral yeas and nays, we are surely no better off if God is alive than if he is dead! In fact, the threat of moral anarchy or nihilism arises from the erosion of belief in God just because the prevailing moral code had been falsely claimed to *derive* from Him epistemologically (via revelation), juridically (in the form of divine commandments), and motivationally (from the love or fear of God)!

Evidently, Kristol's echoing of Nietzsche's challenge backfires: The bite of the challenge is injurious to the religious, rather than to the secular construal of morality.

3. It is a commonplace that "moral codes evolve from the moral experience of communities." But this genesis does not warrant Kristol's normative and motivational view that such codes "can claim authority over behavior only to the degree that individuals are reared to look respectfully, even reverentially, on the moral traditions of their fore-fathers." Surely we ought to winnow the wheat from the chaff in a critical scrutiny of these traditions.

But how, for example, does Kristol's ethical traditionalism enable him to *avoid* asking Jews nowadays to look reverentially at the fact that, at the time of biblical Judaism, women—but not men—were stoned to death for adultery, and that the conditions for obtaining a divorce were brutally asymmetrical as between women and men? Political pressure from rabbinical theocrats in Israel has made it impossible for a Jew there nowadays to get a license to marry a Christian.[23] How does Kristol's conservative stance allow him to erect safeguards against such totalitarian tyranny?

Again, are present-day Christians to show respect for the fact that other devout Western Christians performed barbaric clitoridectomies in the nineteenth century to suppress the sexuality of young girls? Or are they to feel pious stirrings on learning that, with the clergy on his side, Christopher Columbus could see the holy purpose of initiating slave-trading against the people of the Americas, saying, "Let us in the name of the Holy Trinity go on sending all the slaves to Europe that can be sold. The eternal God, our Lord, gives victory to those who follow his way over apparent impossibilities"?[24]

If Kristol were to reply that respect for the repository of ancestral injunctions has to be *selective,* the retort is the one that Sidney Hook gave to Solzhenitsyn: "What besides the methods of reason and intelligence can enable us to make the proper choice between [among] them?"[25] It seems inescapable that *all* traditional ethical injunctions should be subjected to critical scrutiny and distillation.

Kristol's formula founders on the neglect of the precept afforded by Socrates's insight in the *Euthyphro:* If divinely hallowed injunctions are deserving of adoption, then *we* must be the ones—in every epoch anew—to find them worthy. And our only means for doing so are our intelligence and our informed feelings. We have nowhere else to go. Yet Kristol concludes that, since our society no longer defers un-critically or even mindlessly to clerical edicts, contemporary parents are "impotent before such questions" as "What moral instruction should we convey to our children?"

Kristol's application of his traditionalism to contemporary morality features his endorsement, as ancestral divine wisdom, of the inhumane homophobia of biblical Judaism. Referring to the demise of the pro-hibition of homosexuality as "moral disarray," he says mournfully:

> Reform Judaism has even legitimated homosexuality as "an alternative lifestyle," and some Conservative Jews are trying desperately to figure out why they should not go along. The biblical prohibition, which is unequivocal, is no longer powerful enough to withstand the "why not?" of secular-humanist inquiry.[26]

So much the better for the moral challenge from secular humanism, which produced a humane advance over barbarism and cynical hypocrisy.

But, in Kristol's view, the inability of secular humanism to deliver a "compelling, self-justifying moral code," which he employs as red herring, is only the first of its "two fundamental flaws." He reserved his supposed *coup de grâce* for the second:

> A second flaw in secular humanism is even more fundamental, since it is the source of a spiritual disarray that is at the root of moral chaos. If there is one indisputable fact about the human condition it is that no community can survive if it is persuaded—or even if it suspects—that its members are leading meaningless lives in a mean-ingless universe. . . . Secular humanism is brain dead even as its heart continues to pump energy into all of our institutions.

But why can secular humanists not lead richly meaningful lives, just because, in their view, the values of life lie *within* human experience itself? How would our lives be more meaningful, if we did suppose narcissistically that man is the centerpiece of an avowedly inscrutable overall divine purpose, which constitutes *the* meaning of our lives but must remain unknown to our finite minds? Being at the focus of elusive cosmic "meaning" is clearly irrelevant to finding value on this earth: experiencing the embrace of someone we love, the intellectual or artistic life, the fragrance of a rose, the satisfactions of work and friendship, the sounds of music, the panorama of a glorious sunrise or sunset, the biological pleasures of the body, and the delights of wit and humor.

In the movie *Limelight*, Charlie Chaplin put in a nutshell what is wrong with the narcissistic delusion that there is such a thing as *the* meaning of life: Life, said Chaplin, is not a meaning, but a desire. Yet Václav Havel, who has a penchant for mysticism, lists "the meaning of our being" as a basic human question.[27] And a rabbi demands an "ultimate meaning" in human life: "In the atheistic premise, there is no ultimate meaning to human life. It is just there. Now, no human being behaves as if life had no meaning."[28]

As secular humanists see it, there are as many "meanings" as there are fulfillments of human aspirations. It is sheer fantasy, if not arrogance, on the part of theists to proclaim inveterately that their lives must be more meaningful to them than atheists and secular humanists find their own lives to be to themselves. Where is the statistical evidence that despair, depression, suicide, aimlessness, or other dysphoria are more common among unbelievers than among believers? Yet, Kristol insists: "It is crucial to the lives of all citizens, as it is to all human beings at all times, that they encounter a world that possesses a transcendent meaning, a world in which the human experience makes sense."[29]

Regrettably, Kristol did not come to grips with the arguments in Albert Einstein's paper on "Science & Religion," which was delivered in 1941 at the Jewish Theological Seminary in New York.[30] There Einstein first points out: "Nobody, certainly, will deny that the idea of the existence of an omnipotent, just and omnibeneficent personal God is able to accord men solace, help and guidance; also by virtue of its simplicity the concept is accessible to the most undeveloped mind."[31] But then

Einstein issues his cardinal plea, which clashes head-on with Kristol's nostrum: "In their struggle for the ethical good, teachers of religion must have the stature to give up the doctrine of a personal God, that is, give up that source of fear and hope which in the past placed such vast power in the hands of priests."[32]

This rejection of theism as part of Einstein's further denial of supernatural causes impugns Sir Hermann Bondi's reading that Einstein championed a belief in a superintelligence who was the "architect" of the world's complexity.[33] Yet, Bondi himself is staunchly anti-religious.

It is true, if trite, that, if there is deep and widespread demoralization in a community, as well as pervasive disaffection with its institutions, its socio-political organization will crumble, and it will become highly vulnerable to its enemies. Kristol transforms this commonplace into an ominous charge against secular humanism:

> If there is one indisputable fact about the human condition it is that no community can survive if it is persuaded—or even if it suspects— that its members are leading meaningless lives in a meaningless universe.

But the supposition that the godless lead meaningless lives is just an ideological phantasm.

In an article entitled "Can Atheists Be Good Citizens?"[34] Richard John Neuhaus argues for a *negative* answer to the question posed in its title.

First he tries to cope with the fact that Sidney Hook, a lifelong, ardent secular humanist, was a dedicated, fearless critic of totalitarianism for decades who received the Medal of Freedom from the United States government. In that attempt, Neuhaus falls into a confusion between the semantic content of a doctrine with the degree of epistemological confidence that a given supporter of the doctrine may have in it. The content of theism is the assertion that there is a personal God with specified attributes, while the content of atheism is the denial of that claim. But neither content tells us with what degree of confidence a given proponent avows the given tenet.

The Roman Catholic church claims absolute dogmatic, irrevocable certitude for its theism, while Madalyn Murray O'Hair has proclaimed her atheism just as irrevocably. Alternatively, both theism and atheism alike can be espoused with varying degrees of epistemological confidence. Some may regard their belief as a highly probable hypothesis in the light of the evidence, while others may see it more tentatively as the best available working hypothesis.

Theoretical beliefs, however well supported by known evidence, are still fallible or revocable, because of potentially adverse future evidence. It is therefore the better part of wisdom to stop short of espousing one's hypotheses as irrevocably established. Thus, Sidney Hook, Freud, Einstein, and Bertrand Russell, among others, adopted this less-than-dogmatic attitude toward their belief in atheism, but without tampering with its semantic content. Notably, their lack of dogmatism did not, however, constitute a watering down of their atheism into the different doctrine called "agnosticism."

In its technical meaning, agnosticism does not rule out either theism or atheism: It pointedly makes no claim as to the existence of God one way or the other, even tentatively, because it regards the question as unanswerable in principle. Thus, neither theists nor atheists are agnostics. And atheists disavow agnosticism no less than theists do.

This fact was untutoredly overlooked by Robert Bork during his unsuccessful confirmation hearings to become a U.S. Supreme Court Justice. Eyes flashing, Bork told the senators that he was not an agnostic, presumably to convey that he was not irreligious. But Bork's rejection of agnosticism does not rule out his being an atheist.

Neuhaus[35] denies that Sidney Hook was an atheist, claiming that, instead, Hook was an agnostic. Having wrongly assumed that atheism must be *irrevocably* declared true by its champions, Neuhaus concluded that, since Hook was a *fallibilist,* his rejection of theism must be tantamount to agnosticism after all. But this conclusion is false: Hook's commitment was to atheism, not to agnosticism.

Neuhaus is also led to claim incorrectly[36] that the Enlightenment rationalists were "committed to undoubtable certainty," merely because they were atheists. No wonder that, apropos of Laplace's statement to Napoleon, saying that he saw no need for the "hypothesis" of God, Neuhaus declared sorrowfully: "When God has become a hypothesis, we have traveled a very long way from both the gods of the ancient city and the God of the Bible."[37] But why is that deplorable, if modern knowledge compels the demythologizing of the Bible, as indeed it does?

The principal thesis of Neuhaus's article is that atheists cannot be good citizens. Therefore, Hook's actual atheism commits Neuhaus willy-nilly to the further conclusion that Hook is *philosophically* unfit to be a good citizen. Neuhaus's central argument, no less than Kristol's, turns out to run afoul of the moral sterility of theism. And this ethical infertility undermines his attack on the separation of church and state, as well as his irate indictment of those religious people who support that separation.

Yet in his castigation of religious believers who support the organization Americans United for Separation of Church and State, whom he charges with "political atheism," he abjures even the notion of the "existence" of God as too this-worldly! Indeed, we learn: "The transcendent, the ineffable, the totally other, the God who acts in history was tamed and domesticated in order to meet the philosopher's job description for the post of God."[38] But this jeremiad boomerangs: If God is indeed so totally transcendent as to be ineffable, and if He eludes all intelligibility by being "totally other," how can there possibly be any meaning in the *causal* assertion that He "acts in history"?

Indeed, how can we possibly escape the conclusion that talk about such an *avowedly* "totally other" entity is just pretentious babble? Is the insistence on engaging in such discourse not a case of thought pathology, abetted by the penchant to abuse language? If, as we were told in the same vein, Yahweh—the God of Moses—was "above naming and beyond understanding," how can such an entity be *intelligibly* taken on faith even without evidence, let alone be loved or feared?

It is rank political coerciveness for Neuhaus to tell us that, unless we are willing to parrot such gibberish, we are poor citizens. In striking contrast, in a letter written in 1790, George Washington explained to a Jewish community leader in Newport, Rhode Island:

> The citizens of the United States of America . . . all possess alike liberty of conscience and immunities of citizenship. It is now no more that toleration is spoken of, as if it was by the indulgence of one class of people that another enjoyed the exercise of their inherent natural

rights. For happily the government of the United States, which gives to bigotry no sanction, to persecution no assistance, requires only that they who live under its protection, should demean themselves as good citizens, in giving it on all occasions their effectual support.[39]

Bernard Lewis articulates George Washington's distinction between mere toleration and genuine coexistence very well:

In these words, the first president of the United States expressed with striking clarity the real difference between tolerance and coexistence. Tolerance means that a dominant group, whether defined by faith or race or other criteria, allows members of other groups some—but rarely if ever all—of the rights and privileges enjoyed by its own members. Coexistence means equality between the different groups composing a political society as an inherent natural right of all of them—to grant it is no merit, to withhold or limit it is an offense.[40]

Yet, significantly, Neuhaus deploys his charge of poor citizenship even against those believers who have felt driven to take intellectual account of post-Enlightenment developments in the modern world. In fact, he levels the charge of deicide against them.[41]

But the gravamen of Neuhaus's case is yet to come. Having omitted mention of the *fallibilist* kind of atheism held by such secular humanists as Sidney Hook, Neuhaus tendentiously enumerates the doctrines of much less reasonable atheists, and then he asks rhetorically:

Can these atheists be good citizens? It depends, I suppose, on what is meant by good citizenship. We may safely assume that the great majority of these people abide by the laws, pay their taxes, and may even be congenial and helpful neighbors. But can a person who does not acknowledge that he is accountable to a truth higher than the self, external to the self, really be trusted? Locke and Rousseau, among many other worthies, thought not. However confused their theology, they were sure that the social contract was based upon nature, upon the way the world really is. Rousseau's "civil religion" was apparently itself a social construct, but Locke was convinced that the fear of a higher judgement, even an eternal judgement was essential to citizenship.

It follows that an atheist could not be trusted to be a good citizen, and therefore could not be a citizen at all. Locke is rightly celebrated as a champion of religious toleration, but not of irreligion. "Those are not at all to be tolerated who deny the being of a God," he writes in *A Letter Concerning Toleration.* "Promises, covenants, and oaths, which are the bonds of human society, can have no hold upon an atheist. The taking away of God, though but even in thought, dissolves all." The taking away of God dissolves all. Every text becomes pretext, every interpretation misinterpretation, and every oath a deceit.[42]

Neuhaus offers a red herring in his ambiguous rhetorical question: "But can a person who does not acknowledge that he is accountable to a truth higher than the self, external to the self, really be trusted?" A secular humanist's insistence on the indispensability of reliance on the intelligence of the human species patently does not entail, as Neuhaus would have it, that any one of us is morally accountable only to our own self!

Here, he is trading on the vagueness and ambiguity of the phrase "truth higher than self" to allude to the edicts of purported divine revelation of some sort. Unless he does so, the willingness to acknowledge accountability to a "social contract based on nature—on the way the world really is" obviously does not militate in favor of theism as against secular humanism. Indeed, it is secularism that relies on science to tell us about "the way the world really is."

The statements that Neuhaus then quotes or echoes from John Locke are vitiated by the moral sterility of theism, besides being outrageously false on their face. Indeed, we are being treated to scurrilous demagogy when Neuhaus declares that, in the case of an atheist, "Every text becomes pretext, every interpretation misinterpretation, and every oath a deceit."

Ironically, Neuhaus's invocation of Locke boomerangs: According to Locke, citizenship should not be accorded to Roman Catholics either, because these religious believers owe their ultimate allegiance to the foreign pope, rather than to God. Isn't it odd that Neuhaus, the recent convert to Roman Catholicism, made no mention at all of this highly inconvenient fact?

In an important recent article[43] George Weigel relates and deplores the history of allegations in the United States that an ascending tyrannical "Romanism" or "Papism" poses a threat to the pluralism of American democracy. The burden of his article, however, is a plea against a secularist, anti-transcendentalist polity.

Weigel recounts a major episode that, ironically, is a valuable object-lesson of the dangers run by adopting politically an "absolute" standard of morality on theological grounds:

> It is of moment . . . that classic American Protestant anti-Catholicism in the 19th and early 20th centuries simply took it as self-evident that American democracy required a religious foundation: specifically, a Protestant religious foundation. Absent this, it was widely believed there were but two possible outcomes to the American experiment: a revival of premodern despotism (linked to Rome), or moral anarchy leading, in short order, to political collapse.

Significantly, Weigel adds that none of the advocates of this Protestant anti-Catholicism "ever dreamed of advocating a secular policy in which religion would be ruled out of the public debate."

Thus, by Weigel's own account, it was not a secularized state that generated the anti-Catholic turn he bewails; it was rather the denominational insistence on an absolute, divinely sanctioned morality amid the conflicting theological revelations. Despite ecumenicism, the strife among the gospels seems ineradicable: Witness the recent breakdown of negotiations between the Vatican and the Anglican church, which were to yield an ecumenical composition of their theological differences. Or just contemplate the likelihood that Orthodox Jews will become persuaded of the salvific divinity of Christ![44]

Unaware that his chronicle boomerangs, Weigel concludes by misformulating the clash of ideas between secular humanism and a public policy informed by a religiously transcendent morality. As he would have it, this confrontation (Bismarckian *Kulturkampf*) is "a struggle between those who affirm the classic Jewish and Christian notion of an objective moral order, and those who deny on epistemological grounds that there is any such thing as an 'objective moral norm.' " Having posed the issue in these terms, Weigel speaks conjunctively of "secularism and moral relativism."

But surely the secularist's this-worldly warrant for ethical norms is neutral as between an "objectivist" and a "relativist" construal of their epistemological status. To deny that our moral code has a transcendent religious foundation is not to rule out the objectivity of its secular grounds. Nay, ironically, the cacophony of divergent absolutist revelations is effectively tantamount to *moral relativism* as between the rival religious subcultures.

Alas, Neuhaus and Weigel's gravamen against secular humanism, no less than Kristol's, emerges as a shoddy caricature of the doctrine they attack.

*I am grateful to my colleague Professor Richard Gale, who made several valuable expository suggestions in his comments on the first draft.*

## Notes

1. *Las Vegas Review-Journal,* May 5, 1992, p. 7B.
2. *Time,* March 30, 1992, p. 75.
3. *Time,* December 9, 1991.
4. Ibid., p. 62.
5. Ibid., p. 63, caption.
6. Ibid., p. 65.
7. Ibid.
8. See Christopher P. Tourney's review of *The Creationist Movement in Modern America* by R. A. Eve and F. B. Harrold, *American Scientist* 80 (May-June 1992): 292.
9. *London Times,* May 9.
10. Cited by Michael J. Prival, Washington Society for Humanistic Judaism, *Free Inquiry* (Spring 1988): 3.
11. *Free Inquiry* 7, no. 4 (Fall 1987).
12. Ibid., pp. 29–31.
13. *The Unauthorized Version: Truth and Fiction in the Bible* (New York: Knopf, 1992).
14. R. M. Gale, *On the Nature and Existence of God* (New York: Cambridge University Press, 1991), p. 34.

15. Cf. A. Grünbaum, *Validation in the Clinical Theory of Psychoanalysis* (Madison, Conn.: International Universities Press, 1992), chap. 7: "Psychoanalysis and Theism."

16. "The Future of American Jewry," *Commentary* 92, no. 2 (August 1991): 21–26.

17. Ibid., p. 23.

18. See my "Creation as a Pseudo-Explanation in Current Physical Cosmology," *Erkenntnis 35* (1991): 233–54.

19. "Religion and Science: Paul Davies and John Paul II," *Theological Studies* 51 (1990): 314.

20. For a fuller discussion, see Paul Edwards, "The Dependence of Consciousness on the Brain," in *Immortality,* ed. Paul Edwards (New York: Macmillan Publishing Co., 1992), pp. 292–307.

21. I. Kristol, "The Future of American Jewry," pp. 24–25.

22. *Chronicle of Higher Education,* April 22, 1992.

23. Cf. Ian S. Lastick, *For the Land and the Lord: Jewish Fundamentalism in Israel* (New York: Council on Foreign Relations, 1988).

24. Quoted in Adolf Grünbaum, "The Place of Secular Humanism in Current American Political Culture," *Vital Speeches of the Day* 54, no. 2 (November 1, 1987): 43.

25. *Free Inquiry* 7, no. 4 (Fall 1987): 6.

26. I. Kristol, "The Future of American Jewry," p. 25.

27. "A Dream for Czechoslovakia," *The New York Review of Books,* June 25, 1992, p. 12.

28. Louis Jacobs, *The Book of Jewish Belief* (West Orange, N.J.: Behrman House, 1984), p. 10.

29. *Chronicle of Higher Education,* April 22, 1992.

30. Reprinted in D. J. Bronstein and H. M. Schulweis, eds. *Approaches to the Philosophy of Religion* (New York: Prentice Hall, 1954), pp. 68–72.

31. Ibid., p. 70.

32. Ibid., p. 71.

33. "Humanism—The Only Valid Foundation of Ethics," 67th Conway Memorial Lecture, January 24, 1992, London: South Place Ethical Society.

34. *First Things* (Aug./Sept. 1991): 17–21.

35. Ibid., p. 17.

36. Ibid., p. 20.

37. Ibid., p. 18.

38. Ibid.

39. Quoted in Bernard Lewis, "Muslims, Christians, and Jews: The Dream of Coexistence," *New York Review of Books,* March 26, 1992, p. 49.

40. Ibid.

41. R. J. Neuhaus, "Can Atheists Be Good Citizens?" p. 18.

42. Ibid., p. 20.

43. George Weigel, "The New Anti-Catholicism," *Commentary* (June 1992): 25-31.

44. Cf. W. Leibowitz, *Judaism, Human Values and the Jewish State* (Cambridge, Mass.: Harvard University Press, 1992).

# Part II

# Scientific Issues

# 8

# Science, Humanism, and the New Enlightenment

## Vern L. Bullough

This is an attempt to tie together science, humanism, and the new Enlightenment. I have elected to do this in a more personal way to bring my points home. When I began my academic career in the early 1950s, one of the issues in the history of science was the debate over explanations of causes of the scientific revolution, defined at that time as the change from the geocentric to the heliocentric universe. I spent several years pondering the problem and eventually compiled a book on the issue.[1] The long dominance of the geocentric theory demonstrates just how strong a hold presuppositions often have on us and how important it is to recognize that they do.

Though the Greeks had briefly toyed with the idea that the sun might really be the center of the universe, it was the geocentric view, as adopted by Aristotle, that dominated western thought. The issue was not a matter simply of observation since observations were not accurate enough to prove one view more valid than the other, but it was the assumptions and explanations tied in with the geocentric view that gave it dominance. Much of ancient science came to be tied to the geocentric view. An arrow or an object tossed into the air fell to the ground because the earth, a material thing, was at the center and every material thing wanted to come home. Fire rose because it aimed toward the sphere of fire which was above that of earth, water, and air. Seemingly every occurrence could be drawn into such an theory.

In the thirteenth century, St. Thomas Aquinas went further than Aristotle and tied the geocentric theory to Christian theology and made it the cornerstone of Christian faith. At the same time he was doing so, developments in impetus physics and in astronomical observations brought about an ever more complicated geocentric theory far beyond the simple universe that Aristotle had conceived. Copernicus in the sixteenth century, intrigued by the fact that some Greeks had for a time played with the idea of a heliocentric universe, wondered how they could have done so and his treatise on the possibility was published posthumously in 1543.

His new theory, using much of the same data as the geocentric theory, continued to circulate even though it was rejected by the scientists of the time. Instead they continued to develop ever more complicated theories about spheres and counter spheres of motion in order to prop up the geocentric theory. It took a new kind of instrument, the telescope, which raised a number of new contradictions to the geocentric theory, such as the moons of Jupiter and the phases of Venus, before someone, in this case Galileo, had enough courage to challenge the traditional view. Galileo, however, was forced to recant by the inquisition, but despite his recantation the evidence grew and the heliocentric view was consolidated with Newton's theory of gravity.

Still, there was opposition, mainly in Catholic Christianity, which not only canonized St. Thomas Aquinas, but accepted his Christianized Aristotle as the basis of Catholic theology. It was not until the twentieth century that Galileo was finally rehabilitated by Pope John Paul II. Eventually of course the sun was shown to be part of the milky way system and though it might be the center of our solar system, it is not now regarded as a particularly important star in terms of the universe of stars.[2]

I think this account has important lessons for humanists and their reliance on science. Let me amplify. Thomas Kuhn, who also looked at changes in science, explained that while the development of science was incremental, there were also periodic shifts in insights necessary to establish a new paradigm. These shifts resulted when the data, no matter how it was interpreted or squeezed within the existing paradigm, became increasingly unable to explain various anomalies, and this eventually led to a new kind of commitments, a shift in paradigms to new theories.[3] The new theory, however, does not throw out the old data which been empirically verified, but rather uses it to verify the new theory, and it is the willingness of science to eventually follow where the data leads that is the key. Though this view of science has been used by many postmodernists to denigrate scientific truths, arguing instead for everything being relative to the point of view from which one approaches the data, this was not the way Kuhn envisioned it, nor is it the assumption of this writer.

What does seem obvious, however, is that science does not demonstrate the ultimate truth but rather a method, a rational method if you will, of organizing data and empirically verifying it. It is the method of science that is important and this is a point sometimes missed by the critics of science or for that matter of secular humanism. Still the critics have a point. What most such critics are objecting to is the early positivist view which grew out of the Enlightenment and which might best be exemplified by the beliefs of the French philosopher, August Comte (1789–1847), the founder of sociology. Comte proclaimed that all thought evolved through three stages: the theological when men represented the natural phenomena as the products of supernatural agencies; the metaphysical when supernatural forces are replaced by abstract ones; and the positive when humans, eschewing the search for inner natures and essential causes, employ their powers of reason and observation to discern the invariant show of the phenomena, thereby gaining technical mastery over them. Though positivism in its formulation by David Hume (1711–1776) distinguished between analytical knowledge and other forms, in its most extreme form it held that the only valid form of nonanalytic knowledge is scientific. By that it was meant that it had to be testable or verifiable. Propositions could only be accepted as meaningful if some empirical evidence could count for or against them.

This is perhaps asking more of science than it can do, and errors can be made in the name of science if it is pushed too far. When it is pushed too far, the deconstructionists and postmodernists have a basis for criticism by showing that all of us are prisoners of our own culture and civilization.[4] Positivism in its preoccupation with the observable and manipulable, it is argued, reflects a form of technical-scientific practice embodying only one limited human interest. This to me is not only a criticism of some of the extremes of science but an even more severe criticism of traditional religion. Granted that humanists cannot know with absolute certainty that what we advocate is the only true way, few humanists have ever believed that there was. Still it is a cautionary warning.

Presuppositions do play an important role in science, but it is impossible to have a scientific "view from nowhere." To say one's perspective is unaffected by any assumptions and that it is the one correct view of the world is to play God, or do what Donna Haraway called the "God trick," and no scientist can do this.[5] Moreover science does not work this way. What is important in science is that presuppositions or theories change in response to the data, although such changes are not easy, as indicated by my illustration of the explanations of the scientific revolution. In fact there is a general tendency to stick with the same old assumptions, fitting new data into these assumptions until they became so modified and complicated that a new explanation seems more valid.

Still, as Karl Popper pointed out in the 1950s and 1960s, nonscientific propositions might well be meaningful, but in order to determine this we still need to establish some form of verification, and he proposed the concept of falsifiability as a criterion of demarcation. He held that since universal hypotheses are much more readily exposed as false than established as true, scientific activity is prudently directed not against the perpetration of mistakes but against their perpetuation. Thus conjectures should be subjected to the severest imaginable empirically developed tests, in the hope that, if false, they will reveal their falsity. Popper ends up arguing that although science may not ever attain truth, whatever that might be, perhaps it may approach it by erasing errors and increasing in verisimilitude, i.e., approaching the whole truth.[6] Science, or better yet, the methods of science, gives us a way of achieving what might be called "objective knowledge" because it is not an individual product but rather the product of a community of scientists negotiating with each other. Results of data gathering and observations are published and judged by one's peers, and this means that objectivity in science consists of those views that seem to be agreed upon by the scientific community at any one point in time, and these change with new data or new assumptions.

This, however, is the view of a historian of science in the mid-1990s. Looking back over the past hundred years we find that the advocates of science have not always been so conscious of their own limited prisms, and have assumed that only they had the "truth." On these so-called truths they often erected new theoretical bases, or in sociological terms, "social constructions" of how society should be, based upon their belief that science proved their point of view. Some of these system makers proved every bit as dogmatic as their theological predecessors had been. Moreover, they often tried to impose their views on society.

A good illustration of this comes from my research into human sexuality, where the lessons to learn about the dangers of dogmatic reliance upon science are everywhere apparent. Let me look at just one area, menstruation. During the last part of the nineteenth century American physicians explored several theories of menstruation since there seemed to be no simple explanation as to why human females went through such a process. In general, the Americans were aware of the theories of John Power of London who postulated that ovulation and menstruation were connected,[7] although there was uncertainty as to whether the onset of the menses marked ovulation or whether it came afterwards. Even as late as the 1890s, when the first experimental work leading to the understanding of human hormones was taking place, American physicians were still discussing the question of whether the ovaries triggered menstruation, whether the uterus was an independent organ and performed the menstrual function without external aid, or whether the fallopian tubes were responsible for the monthly flow.[8]

Some physicians, perhaps influenced by the Victorian disgust at the sexual and reproductive process, considered menstruation a pathological condition. Such physicians mixed what might be called Christian fundamentalism with inadequate data and claimed it as science. They taught that in Paradise humans had reproduced asexually and that it was only when man had fallen that perfection had been replaced by the evil of sex. Menstruation was a sign of this curse of God. An 1875 article in the *American Journal of Obstetrics,* for example, following this view of menstruation as a curse, argued that menstruation was pathological, proof of the inactivity and threatened atrophy of the uterus. As evidence of its unnaturalness the author claimed that conception was most likely when intercourse occurred during the monthly flow but intercourse at such times was dangerous and forbidden because the menstrual blood was the source of male gonorrhea. Since menstruation therefore stood in the way of fruitful coitus it obviously had not been ordained by nature.[9] Though there are many unexamined assumptions in such arguments, it is typical of what passed for research among many physicians of the time.

In 1861 E. F. W. Pflüger demonstrated that menstruation did not take place in women whose ovaries had been removed, a finding which reinforced the ovarian theory but did not end the debate over the physiology of menstruation since Pflüger hypothesized that there was a mechanical stimulus of nerves by the growing follicle which was responsible for congestion and menstrual bleeding. This led him to argue that menstruation and ovulation occurred simultaneously.[10] It was not until the twentieth century and a better understanding of the hormonal process that the timing of ovulation and stimulus involved was fully understood. In the meantime, Pflüger's theory that nervous stimulation triggered menstruation was widely accepted.

The theory itself was not unreasonable based upon observations of the influence of stress and tension on menstrual irregularity. But many observers were not content simply to see this as one factor in the onset of the menses. Instead they used it to erect new theories about the nature and purpose of the female. From an observable and testable fact, they erected a theory which they could not test. Leaders in the theorizing were those males who seemed to be most threatened by the changing relationships between men and women that were taking place in the nineteenth century. One indicator of this change was the demand by women to enter colleges and universities. In response to this female academies and seminaries in the United States began to multiply and encouraged by this some (both males and females) demanded still greater educational opportunities for women. The first signs of success in the campaign came in 1833 when Oberlin College was established as a coeducational institution. Similar demands and changes were occurring in other parts of the western world and as a new generation of educated women emerged, they sought to enter the all-male professions. Medical schools found themselves under attack for failure to admit women, and a few women such as Elizabeth Blackwell managed to receive medical training, something that the rank and file of the medical profession opposed, some of them with great hostility.

Typical of the opposition to this intrusion of women into a male profession was Edward H. Clarke, a physician at Harvard medical college. His argument against the admission of women to the medical college was based on his concept of menstrual disability, which to his mind was objective, rational, and verifiable. The argument was also advanced by several other physicians, but Clarke elevated it to a whole theory of womanhood. In 1873 he wrote that though women undoubtedly had the right to do anything of which they were physically capable, the physiology of being female put limits on their opportunities. One of the things they could not do, and still retain their good health, was to be educated in the pattern and model of men adopted for men. Clarke explained this by stating that while the male developed steadily and gradually from birth to manhood, the female, at puberty, had a sudden and unique period of growth when the development of the reproductive system took place.[11] This, he said, following Pflüger, involved special demands upon the female nervous system since it had to work on developing not only the brain, as it did in males, but also the reproductive organs. This made the female different from the male whose nervous system could concentrate solely on intellectual development. He then argued that since the nervous system could not do "two things well at the same time," it was important for the female between the ages of twelve and twenty to concentrate most of her energy on developing her reproductive system. This implied that females should not devote much time to higher education because if they did so the signals from the developing organs of reproduction would be ignored in favor of those coming from the overactive brain.

There was more to his argument than this brief summary but he eventually concluded that women who concentrated upon education rather than the development of their reproductive system underwent mental changes. Not possessing the physical attributes of a man, they tended to lose the "maternal instincts" of a woman and become coarse and forceful, with the result that a new class of sexless humans analogous to eunuchs was appearing among women. To solve this alarming problem, he recommended strict separation of the sexes during education, particularly after elementary school. He urged that female schools provide periodic rest periods for students during their menstrual periods. The young women would also have shorter study periods since they were by nature weak and less able to cope.

> A girl cannot spend more than four, or, in occasional instances, five hours of force daily upon her studies, and leave sufficient margin for the general physical growth that she must make. . . . If she puts as much force into her brain education as a boy, the brain or the special apparatus (i.e., the reproductive system) will suffer.[12]

Inevitably, he "reluctantly" concluded that women could not be admitted to Harvard Medical School or any intellectual competition with males not only for their own protection but for the preservation of the human race.

Though it would seem obvious to the reader of today that Clarke was using his own biases and prejudices to construct a theory of female inferiority, it was not so evident to his contemporaries who subscribed to the dispassionate truth of science.[13] Even those physicians who challenged the association of menstruation with development of the nervous system, simply replaced it with other theories emphasizing female inferiority and instability. This was the case of John Goodman's theory of "menstrual wave,"[14] which was utilized by George J. Englemann in his presidential address before the American Gynecological Society in 1900 to urge that schools for girls should heed the "instability and susceptibility of the girl during the functional waves which permeate her entire being," by providing rest during the menstrual periods.[15]

Clarke and others were regarded as scientists, and their beliefs had the imprimatur of science. Even those educational institutions that admitted women wanted to make sure that women retained their femininity. The University of Wisconsin Board of Regents held in 1877 that

> every physiologist is well aware that at stated times, nature makes
> a great demand upon the energies of early womanhood and that at
> these times great caution must be exercised lest injury be done.

Though education for women was greatly to be desired,

> it is better that the future matrons of the state should be without
> a University training than that it should be produced at the fearful
> expense of ruined health; better that the future mothers of the state
> should be robust, hearty, healthy women, than that, by over study,
> they entail upon their descendants the germs of disease.[16]

## Conclusion

The illustration of the political conclusions drawn from assumptions about menstruation is important to keep in mind in any discussion of science and humanism. It is a lesson that the deconstructionists and postmodernists have emphasized and one we need to recognize. We have to continually re-examine our assumptions, not only in regard to major theories such as the geocentric or heliocentric universe, but also in our everyday life. What we proclaim to be the "scientific truth" is often overlaid with existing prejudices, beliefs, and superstitions, which in spite of scientific data do not easily change. This was true in the past and is still true today. New findings are usually incorporated into old paradigms until the weight of the contradictions finally forces a change. But these changes in paradigm are also often influenced by other factors. Many of these are political.

Where does this leave humanists? I think it emphasizes the importance of skepticism. We cannot afford to be dogmatic. The deists of the eighteenth and early nineteenth century adopted a world view based on the Enlightenment view of science, which put God as sort of a benevolent clock winder, but deism itself disappeared as a significant movement when science itself progressed beyond the simplistic assumptions of the eighteenth century. Without claiming to have the final answer, we have to put our beliefs to the test more or less constantly. We also have to test the popular assumptions of our time. We might not always be able to disprove them but we can certainly indicate a high probability that they are wrong. What we cannot do, however, is to be stuck in the assumptions of the eighteenth-century Enlightenment. We have to adjust and recognize that science is something humans have developed. It is the best method we have, but it is not infallible because it depends on our assumptions, assumptions that are continually subject to change as new data comes in. The humanist then becomes the eternal seeker after the truth, knowing that we do not have the final answers, but certain that we have the best method yet devised for determining the truth or falsity of most of today's assumptions. In short, we need to adopt the principles of the new Enlightenment, conscious of our past errors, but continually willing to explore and adjust to new challenges. Ultimately even Pope John Paul II adjusted to the reality of Galileo's findings. Humanists at least know that they are the vanguard of the future. Personally I find the whole process of being a humanist a challenging and interesting experience.

## Notes

1. Vern L. Bullough, *The Scientific Revolution* (New York: Holt, Rinehart and Winston, 1969; Huntington Park, N.Y.: Robert E. Krieger, 1978).

2. For an overview of these developments, see ibid.

3. Thomas S. Kuhn, *The Structure of Scientific Revolutions* (Chicago: University of Chicago Press, 1962).

4. J. Habermas, *Knowledge and Human Interests* (London: 1972); Michel Foucault, *Power/Knowledge: Selected Interviews and Other Writings, 1972-77* (New York: Pantheon, 1981); Huber L. Dreyfus and Paul Rabinoe, *Michel Foucault: Beyond Structuralism and Hermeneutics,* 2d ed. (Chicago: University of Chicago Press, 1983).

5. Donna Haraway, "Situated Knowledges: The Science Question in Feminism and the Privilege of Partial Perspective," *Feminist Studies* 14 (1988): 575-99.

6. Karl Popper, *The Logic of Scientific Discovery* (London: Hutchinson, 1959), and *Conjectures and Refutations* (London: Routledge and Kegan Paul, 1963).

7. John Power, *Essays on the Female Economy* (London: Burgess and Hill, 1831); G. F. Gridwood, "Theory of Menstruation," *Lancet*, 1842–43, i: 825–30; J. Bennet, "On Healthy and Morbid Menstruation," *Lancet*, 1852, i: 825–30.

8. M. M. Smith, "Menstruation and Some of Its Effects Upon the Normal Mentalization of Woman," *Memphis Medical Monthly* 16 (1896): 393–99; Frederick Fluhman, *Menstrual Disorders, Diagnosis and Treatment* (Philadelphia: W. B. Saunders, 1939).

9. A. F. A. King, "A New Basis for Uterine Pathology," *American Journal of Obstetrics* 8 (1875–76): 242–43.

10. E. F. W. Pflüger, *Ueber die Eierstöcke der Sügethiere und des Menschen* (Leipzig: Engelmann, 1863).

11. Edward H. Clarke, *Sex in Education; or a Fair Chance for Girls* (Boston: James R. Osgood & Co., 1873), pp. 37–38.

12. Ibid., 156–57.

13. For further amplification of his views see Vern L. Bullough and Martha L. Voght, "Women, Menstruation, and Nineteenth-Century Medicine," *Bulletin of the History of Medicine* 47 (1973): 66–82.

14. John Goodman, "The Menstrual Cycle," *Transactions* (American Gynecological Society) 2 (1877): 650–62; "The cyclical theory of menstruation," *American Journal of Obstetrics* 11 (1878): 673–94.

15. George W. Englemann, "The American Girl of Today: The Influence of Modern Education on Functional Development," *Transactions* (American Gynecological Society) 25 (1900): 8–45.

16. Board of Regents, University of Wisconsin, *Annual Report for the Year Ending, September 30, 1877* (Madison: 1877), 45.

# 9

# From Aristotle to the New Age

## Jean-Claude Pecker

Has History (with a capital H) reached its very end? This question has recently been posed by Francis Fukuyama.[1] At first, it seemed to many commentators a plea for the universal and necessary tendency toward a market economy, more than an objective statement of what is actually happening. In essence, and to use a scientific metaphor borrowed from the realm of classical physics, the world entropy is now coming near to reaching a constant value, its maximum possible value. Mankind is behaving more or less like a closed vessel (see figure 1) in which a little liquid ether is left to vaporize. The isolated system tends always to homogeneity. One could also say that order is leading to disorder. That is, in essence, the "second principle of thermodynamics," the one which imposes on our mind the irresistible idea of the "arrow of time," an idea definitely linked with that of "irreversibility," the compelling trend toward homogeneity in any *isolated* system. I would have been tempted to write also a trend "from order to chaos," would it not have been for the strong warning[2] that chaos, strictly speaking, implies a structure, and can even generate some other structures; it does not imply a maximum of entropy or actually the end of anything! Perhaps History is indeed closer to a pregnant state of chaos, but certainly not close

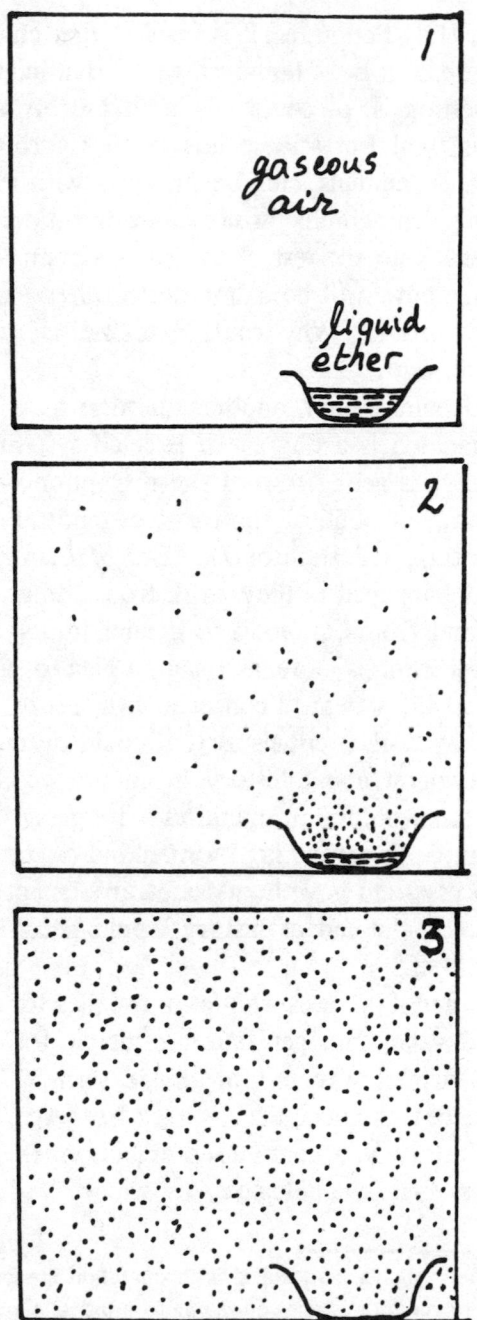

Fig. 1: Evolution of an isolated system—from "order" to "disorder," toward "equilibrium."

to its end. In the Holy Scriptures, it is implied that chaos is at the beginning, not at the end. It is the homogenization that indeed is *the* end!

The first question is, of course, Is mankind an isolated system? Roughly speaking, yes! But science has to be rigorous. Man indeed interacts with plants, animals, etc.; he interacts with the whole earth, if only through his demographic growth and evolution. Moreover, the earth itself interacts with the rest of the solar system. The end of the earth, so we think now, will be a hot death, after vaporization, and we are still very far from it. Why would mankind have already reached any state of equilibrium?

To make this point clearer, another question must be asked: Even assuming that mankind is acting as an isolated system, can we really accept that no force is acting toward the disruption of economies, of sociological patterns, of political contrasts, of cultural disparities? Unfortunately, since the publication of *The End of History, and the Last Man,* enough has happened in the world, from Moldavia to Somalia, from Peru to Ireland, from Cambodia to Bosnia, for us to believe indeed that the book in question is more, as I said, a plea for liberal economy, than a sincere and valid statement concerning the history of mankind.

Aside from any explicit criticisms that could be made against this false entropic viewpoint about history, it cannot be denied that one of the factors of the evolution of mankind is progress* in our scientific knowledge of nature. The mastering by mankind of the physical forces, of the natural processes, etc. is, without doubt, an element of our historical development. Hence, the end of history would necessitate and imply an end of science itself.

In that sense, and if one sees with favor the plea for a liberal market economy to be a valid homogenization principle for the world, one must consider as necessary to *put an end* to science! This is perhaps where the antiscience movement, the so-called New Age, is coming from.

If one sees history as a continuous evolution, then the killing of science may seem somewhat illogical, as science will then disappear

---

*Perhaps "progress" should be qualified with quotation marks, since progress in technology or modern comforts is not equivalent to progress either in ethical values or even in overall average welfare, especially when we consider the masses of the poor.

naturally. If one sees history on the contrary as a series of revolutions, some of which imply a move backward (in freedom, in ethical values, in level of living, in political independence, etc.), then one could very well admit that the existence of scientific revolutions would, in itself, abolish more or less the assumption that science is progressing, and therefore it does not act as a one-way motor of human evolution. One could thus reach the end of history without the need to kill scientific activity. This is a very contradictory attitude, since it is willing at the same time to keep the benefit of science, without letting it be a motor to history. But it may be so: one could then see technocrats hand in hand with gurus!

And this is why it is not inappropriate to discuss the validity of the theories of scientific revolutions, as it was broadly discussed by Thomas Kuhn[3] and, after him, by many historians of science. The theses of Thomas Kuhn are widely known. But Kuhn himself has recognized that they were not always very clear; so I feel his ideas need somewhat to be abstracted, which hereafter, in my own way, I shall do.

A basic notion, in Kuhn, is that of *paradigm*. But, as was clearly pointed out by some of his critics,[4] this is a rather ambiguous word. According to Kuhn, it is to be used with two very different meanings. On the one hand,[5] it is a *whole set* of beliefs, of accepted values, of recognized techniques, common to the members of a "given group." On the other hand, the word denotes only *some* element of *some* sub-set of this complete set. In other words, the word has two meanings. The first one is in essence a "sociological" one; the second one is not yet too clear, in Kuhn's terms: the paradigm is then an "accomplishment" in the past, and can be used as an example. I must say that, after having read Kuhn's reply to criticisms, I am not quite sure that he has indeed cleared all of them. I would like better to stick to an example, a well-known one, treated by Kuhn himself at length, the so-called "Copernican revolution." It is considered as progress, of course, but as a revolution also, implying a radically deep change in "paradigm" in the sociological sense, a word for which I propose to substitute "doctrine." Kuhn's idea is that the Copernican treatise has been the main act in putting in the waste basket the Aristotelian doctrine (I refuse to say paradigm, in this sense, any more).

But, of course, as well known, the Aristotelian doctrine was, first: *explain the phenomena.* Perhaps it is true to say that, at the time of Aristotle, one had not a very clear idea of what was indeed the nature of the phenomena to explain; the notion of experimental science had to wait for Francis Bacon to emerge. Still, Aristotle had a pretty general idea of whatever phenomena he wanted to explain by the introduction of proper paradigms. In other words, paradigms were a set of principles at the basis of a representation of the observed data, however general and badly described they may have been; this was the basis of any "theory of the world." Such a theory was felt as a need; and philosophers of the Aristotelian tradition felt it very acutely indeed.

So the Aristotelian theory implied, in the details, the use of many "paradigms" (in the second sense of the word). One is the *eternity* of the universe. A second and third are the facts that all motions are *circular* and *uniform* in the astral world. This implies a difference *in essence* between the astral and sublunar world, and this is a fourth important paradigm. A fifth one is that there must be, *at the center of any orb,* some *material* object, the Earth or some other celestial body. A sixth one is that the center of the universe should indeed be the *Earth.* Thus, there were at least six paradigms, in the field of astronomy, and several others in physics, or in biology. Did Copernicus put all of these ideas to rest? Certainly, but it happened little by little.

The eternity of the universe was still admitted by the modern neo-positivists, including Einstein. It has been called into question only since the discovery of the so-called expanding universe, and still not unanimously. To quote only one group of authors, let us remember that Hoyle and coworkers[6] have introduced the concept of continuous creation, in order to reconcile expansion, considered as real, and stationarity, which is considered a philosophical necessity.

True enough, Copernicus questioned the fact that Earth was at the center of the universe. But the need for a material mass to be at the center of an orb was already ruled out by Ptolemy. And the non-circularity of motions was shown by the elliptic orbs of Kepler. The uniformity of motion was dispensed with by the introduction of "equans" by Copernicus. The identity, in essence, of the astral and sublunar world was demonstrated in 1574 by Tycho-Brahe, observing a supernova. So the paradigms of Aristotelianism were overturned *progressively* with Ptolemy, Copernicus, Tycho, Kepler, and Hubble et al., not all at once! And so far as the main "doctrine" is concerned, the one designed to "explain the phenomena," it is still valid for the physicists of nature.

Actually, since the Greeks, the Aristotelian doctrine and the Ptolemaïc system have been successively in and out of favor. The Ptolemaïc system violates one of principles of the Aristotelian cosmology, in that the former does not need to have a material mass at the center of the orbs but only a more or less geometric moving point, which is not massive, not material, and neither a planet nor the sun. Therefore, in regard to one of his important paradigms, Aristotle was contradicted by Ptolemy; and the latter was forced to that contradiction because of the need to "explain the phenomena." Hence, Aristotle was eventually contradicted by his own first principle.

This difficulty appeared as obvious to the philosophers of the Cordoba school, Averroes and Maimonides, in the twelfth century. Both of them actually wanted to fight against Ptolemy, not because they thought that the solar system had the sun as its center, but because they preferred to save the paradigm that seemed to them the more important, the one postulating massive centers, and to violate the one that seemed to them less important: the one describing the apparent motions of the planets. By so doing, they explained the Scriptures, too, instead of the phenomena, a Platonic tendency, which has been a permanent part of the whole history of science.

It is interesting to note that neither Averroes nor Maimonides, who both suffered from the Catholic absolutist regime of Ferdinand, were able to create a lasting school of thought; and if they fought against the Ptolemaïc system, it was then left almost unnoticed and was forgotten at the time of Copernicus, not such a long time later.

Words have their weight. Kuhn tries to justify himself and his choice of words, but the very concept of revolution has been misused indeed. Actually, as noticed very subtly by Feuer, the "new" ideas of the "scientific revolutionaries" could emerge because of the active help of senior scientists of the previous generation. The so-called "scientific revolutions" have indeed never been followed by any social movement, even in the universities. I have shown that they were not sudden. In their time, they were not even shocking!

Still, it is clear that many fans of Kuhn are going as far as to imply, by the use of this word "revolution," that one can rewrite science, from scratch, just as it is possible to guillotine a king, to change a constitution, or to legalize a dictatorship. Not only is history conceived as being able to stop: history can even come back to some previous situation, start again, and be rewritten. An obvious misconception! And it may not be inappropriate to remind the reader that some so-called historians go now so far as to negate the Nazis' crimes, and to erase years of crimes in Europe. There were actually Nazis in many European countries, and we cannot really forget this period. History does record its revolutions, but they are real revolutions, with all the accompanying brutality. We see that one can build a new regime on the severed head of Louis XVI or that of Charles I. We see every day the fall of many statues, bronze and stone as well as intellectual statues. First Stalin, Lenin, and Marx disappeared from the official squares and buildings. Recently Enver Hodja departed. Tomorrow who else?

Therefore the very expression "scientific revolution," in its crudity, carries a lot of unpleasant and drastic connotations. In a way, it justifies the notion that, once Copernicus was accepted there, the statue of Ptolemy was torn down. Newton bloomed on top of the somewhat crushed monument. Then Einstein came, and he violently threw out the beautiful statue of Newton, and nothing is left of it.

The continuation of this is clear. Someone will come and throw away the statue of Einstein. Nothing in science is known forever. We can accept as valid all kinds of strong ideas; if they are not valid today, they might be justified later on, and become the pillars of a new revolution, even of a new era in science.

Worse than the shows given by a few cranks, one sees in this tendency the menacing entrance of false sciences, in the name of the possibility of a revolution: The "revolution" of parapsychology; the "revolution" of extraterrestrial life, the extraterrestrial visitors going as far[7] as to change our views in cosmology. The "revolution" in medical science will allow us to cure cancer with homeopathy. One sees at first the extreme danger of taking too literally Kuhn's words. The revolutions of science are not indeed revolutions as they are in history; one should not start from Kuhn's ideas to postulate, with the end of history, the end of science in successive mini-revolutions each one canceling out the preceding.* This type of deviation finds its roots in the careful way that Kuhn, obviously, has not properly analyzed the scientific activity, in its search for describing and modeling the objective reality of nature. It is the insistence on this objective reality that is the basis of modern science.

A certain feeling of freedom vis-à-vis the established domains of science has undoubtedly been strongly reinforced, in the eyes of the ignorant public, by the birth of two domains of science, one called "relativity," the other, quantum mechanics, with its "relations of uncertainty." For us, these words have a very strict and well-defined meaning; for even the cultured layman, it means more or less that "everything is relative" (i.e., not safe, not sure, not established, not absolute, even not "true," and certainly quite provisional, and subject to further revisions), and that everything is "uncertain" (i.e., again, not true, not safe, not established, not absolute, quite provisional, and subject to

---

*This idea can be illustrated in a diagram (figure 2A), which is very rough, of course, but perhaps illustrative. Let us consider the entropy of an *isolated* system, as implied, more or less, by Fukuyama's own view of history. It has to grow and it tends asymptotically toward some asymptotic state of equilibrium, which, according to Fukuyama, we are not far from. Obviously, what we have said leads to describing Kuhn's view in a similar way: The "knowledge" undergoes revolutions, and a lot of previous knowledge can be therefore erased, leading to a true regression in knowledge; this phenomenon is actually true, and is a sociological fact, notably in the United States (figure 2C). Shall we in these conditions reach any sort of equilibrium? Although Kuhn does not say anything about it, we may accept this trend as a possibility; it denies, on the one hand, the very progress of science; on the other hand, it means the insensibility of the social world to this progress. But we do feel that the sociological loss of knowledge (and of confidence in science) has other causes than the scientific revolutions, as described by Kuhn. Therefore, we have a tendency to feel that the progress of science, if not continuous, is affected only rarely and slightly by some loss of previous knowledge, or assumed knowledge (figure 2B). A revolution is more a cumulative progress, which results, at times, in a "reordering" of the bulk of knowledge, followed more or less by a slow digestion, implying elaboration of the reordering in question.

Actually, the loss of confidence in science, the rise of irrational ideas, has many causes, not to be looked for in Kuhn's theory. Some will certainly be mentioned by others.

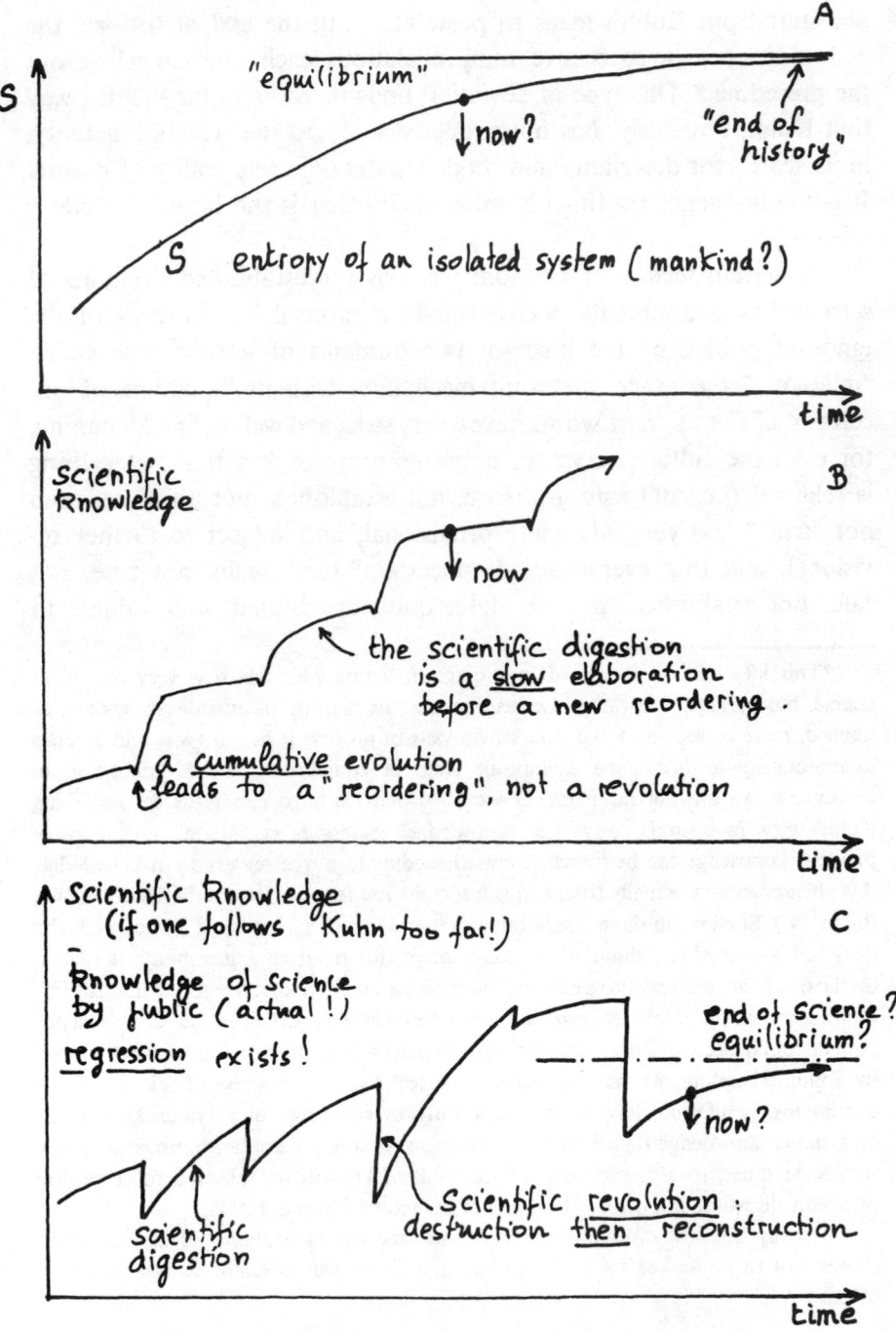

Fig. 2: Science with time

revisions); suggests implicitly an absurd correlation between the behavior of science and scientists and the behavior of elementary particles. It shows once more that words have an enormous weight, and that scientists should perhaps be more careful in introducing, in their language, words that have already a clear signification in everyday life. In any case, and along the same line of thought, inspired by the misuse of words, we should by now recognize that science does not react to its "revolutions" as does history. Fortunately, science does not work that way!

What happens indeed seems to me quite different. At any given time, paradigms, in important number, are present in science. Someone then comes and, on the basis of some new experimental knowledge, suggests a new theory able to encompass some of the preceding paradigms. A clear and recent example is the birth of general relativity; it encompasses special relativity and Newtonian mechanics, but it does not invalidate them; they appear only as special cases. Newtonian gravitation can emerge from the general-relativity equations if one puts in them the velocity of light as infinite; and special relativity corresponds to the case where the gravitational constant G is neglected, and equated to zero. This progress is, at the same time, a simplification of the expression of the physical laws, but a complication is that it shows that no physical constant can be neglected. Incidentally, this progress is the very reason why I have always been suspicious of Friedmann's cosmology, which equates Einstein's cosmological constant $\Lambda$ to zero, in an effort toward simplicity, whenever the nonzero value of $\Lambda$ enables us to account for other paradigms than simply the expansion of the observed universe, such as for example, the quasi-isotropy of the background, the so-called cosmological radiation.

Here I should perhaps introduce an argument against an excessive use of the principle of Occam's razor. In essence, this well-known principle expresses the idea that one should *never,* in a theory, introduce *unnecessary* new entities, unless needed to "explain the phenomena," i.e., *all* the observed phenomena. But one extends it often to the introduction of new parameters, and one then sees that Friedmann's work was inspired by this principle; it was, at his time, unnecessary to introduce the cosmological constant $\Lambda$, as it was, for Newton, unnecessary to assume that the velocity of light $c$ is finite. But one should not go too far.

For example, the concept of God suggests a very minimal set of hypotheses; if one assumes the existence of a God, as creator of nature, one has really reached an ideally simple vision, which gives a reply to every question, and which replaces, for the believers, any other theory, any other hypothesis. Perhaps I am somewhat too blunt in this matter; after all, what I have said justifies, in a way, Occam's principle, but the long-term consequence is to eliminate scientific research. I would prefer instead to forget about Occam's doctrine, a very extreme case of Platonism, and I would like to keep to the old Aristotelian doctrine, i.e., "first, explain the phenomena," i.e., at any expense, account for the whole set of observations. This doctrine was not always applied by Aristotle himself, perhaps because he did not know well some of the phenomena that were better known after Hipparchus, at the time of Ptolemy.

In any case, if we follow the history of ideas throughout the continuous and cumulative progress of science, we see, in the philosophical interpretations of this history, a triangular approach, which has regularly led to conflicts, and never affected, however, the steady progress of a knowledge based upon new observations. The three angles of the triangle are, in essence, the three coexisting doctrines of Aristotle, Plato, and Pythagoras. It seems to me that scientists, all the way through, have been essentially inspired by the Aristotelian principle: "first, explain the phenomena." At the present time, the return of a certain mystical dualism,[8] which claims that there are two ways to knowledge, that of scientific knowledge, and that of complementary mystical intuition, this return is indicative of a neo-Platonism. Along these lines, explaining the phenomena is not the only thing one should have in mind; paradigms should be unified by some religious doctrine. Occam's principle is in that line, with its very clear metaphysical connotation. The third angle of the triangle, that of Pythagoras, is a very frequent temptation, nowadays, for those who do not even consider it important to explain the phenomena, so long as their paradigms can achieve only the building of a beautiful structure, based on numerical coincidences, or on geometrical constructions of an almost purely abstract nature. Sometimes, I feel the modern physics of particles (the superstrings, for example) is more or less following this path, or at least is tempted by the beauty of the construction to consider this the important thing, though it is very far from the observations. But the Aristotelians should keep these efforts under control, one hopes. We have therefore three different "doctrines" which coexist; at times, the dominance of one or the other has acted as a hindrance to the progress of science; the Aristotelian doctrine seems the only one indeed to be compatible with it.

In any case, the announced awakening of the "new age" is, without a doubt, a blooming of the mystical tendencies, which implies either a return of mystical dualism, as a transition phase perhaps, or even a comeback of a completely mystical view, as its ultimate tendency, in the light of the Pythagorean sects, now reborn in many forms.

I hope that one has seen why, in this regrettable evolution, I do consider the influential theory of scientific revolutions to have played an important but negative role, and why I have linked it with the recent discussions related to the so-called end of history.

## Notes

1. F. Fukuyama, *The End of History, and the Last Man* (New York: The Free Press, Macmillan, 1991); French translation, *La fin de l-Histoire et le dernier Homme* (Paris: Flammarion, 1991).
2. Evry Schatzman. Private communication.
3. T. Kuhn, *The Structure of Scientific Revolutions* (Chicago, Ill.: The University of Chicago Press, 1962, revised 1970). Used by the author in its French translation, *La structure des Révolutions Scientifiques* (Paris: Flammarion, 1983).
4. M. Masterman, "The Nature of a Paradigm," in I. Lakatos and A. Musgrave, *Criticism and the Growth of Knowledge* (Cambridge, U.K.: Cambridge University Press, 1970); D. Shapere, "The Structure of Scientific Revolutions," *Philosophical Review* 73 (1964): 384–94.
5. T. Kuhn, *Scientific Revolutions* (1970), postface. The translation of this postface appears in the 1983 Flammarion edition; "Reflexions on My Critics," in I. Lakatos and A. Musgrave, *Criticism and the Growth of Knowledge;* "Second Thoughts on Paradigms," in F. Suppe, *The Structure of Scientific Theories* (Urbana, Ill.: 1970 or 1971) (as quoted by Kuhn in *Scientific Revolutions*).
6. F. Hoyle, H. Bondi, T. Gold, R. Littleton. Read for example: F. Hoyle,

*Month. Not. Roy. Astron.Soc.* 108 (1948): 372, or F. Hoyle, *Steady State Cosomology Revisited* (United Kingdom: Univ. College, Cardiff Press, 1980).

7. J.-P. Petit, *Enquête sur des extra-terrestres* (Paris: Albin Michel, 1991).

8. J.-C. Pecker, "The Return of Mystical Dualism," in *Neofundamentalism,* ed. Paul Kurtz for the Academy of Humanism (Buffalo, N.Y: Prometheus Books, 1988).

# 10

# Scientific Humanism and Religion[*]

## Edward O. Wilson

In 1986 I was invited by the Committee on Human Values of the Roman Catholic Bishops of the United States to be one of four scientists to join them and a group of theologians and scientists from Catholic colleges and universities in a free-wheeling discussion of the relationship between science and religion. It was held in a retreat outside Detroit; there was no audience. The other three scientists were Freeman Dyson, the theoretical physicist from the Institute of Advanced Studies at Princeton; Roger Sperry, the neurobiologist from Cal Tech; and Jerome Lejeune, a French geneticist best known for his discovery that Down's syndrome is caused by a chromosome anomaly. Dyson, Sperry, and I are scientific humanists; Lejeune is a devout Catholic and a member of the Pontifical Academy of Sciences. Also present were about fifteen bishops, including Cardinal Law of Boston and Cardinal (then Archbishop) James Hickey of Washington.

It was an altogether remarkable event. The motivation of the Bishop's Committee appeared to be to somehow accommodate the Church to science, to look for common ground, perhaps to chart areas in the

---

*Originally published in *Free Inquiry* (Spring 1991): 20–23, 56. Reprinted with permission.

Church still closed to free inquiry, and conversely perhaps to find safe and authenticated channels through which to guide the Church to a quieter harbor of more liberal thought. As one theologian put it, "Science left the Church after Aquinas, and we never tried to call her back." What, I had to ask, was happening here?

Our discussion, which covered two full days (September 16 and 17), was frank. Humanistic ideas, of which my own were perhaps the most unequivocal, were politely received and addressed. No subject was implicitly regarded as taboo except abortion, and not even the scientists wanted to put that explosive issue on the table. Scientific challenges were made to the religious view of human meaning. Artificial birth control was advocated, to my surprise, not just by the three humanistic scientists but by several of the bishops.

My approach was uncompromisingly humanistic but conciliatory. I looked for common ground, searching constructively for ways for the Church to liberalize and accommodate science—and in its turn to confront science with hard ethical and ontological questions. What follows is the talk I presented to the Catholic bishops, without a single word changed.

At the Cologne Cathedral, in 1980, Pope John Paul II said that science has added "wings to the spirit of modern awareness." Yet science does not threaten the core of religious belief:

> We have no fear, indeed we regard it as excluded, that a branch of science or branch of knowledge, based on reason and proceeding methodically and securely, can arrive at knowledge that comes into conflict with the truth of faith. This can be the case only where the differentiation between the orders of knowledge is overlooked or denied.

That last sentence, reaffirming Augustine's two books on God, takes us to the heart of the real dialogue between religion and science. Although many theologians and lay philosophers like to deny it, I believe that traditional religious belief and scientific knowledge depict the universe in radically different ways. At bedrock they are incompatible and mutually exclusive. The materialist (or "humanist" or "naturalist") position can be put in a phrase: There is only one book, and it was written in a manner too strange and subtle to be foretold by the prophets and Church fathers.

But, there is another side to the story, one that makes the contrast in world views still more interesting. The materialist position presupposes no final answers. It is an undeniable fact that faith is in our bones, that religious belief is a part of human nature and seemingly vital to social existence. Take away one faith, and another rushes in to fill the void. Take away that, and some secular equivalent such as Marxism intrudes, replete with sacred texts and icons. Take away all these faiths and rely wholly on skepticism and personal inquiry—if you can—and the fabric of society would likely start to unravel. This phenomenon, so strange and subtle as to daunt materialist explanation, is in my opinion the most promising focus for a dialogue between theologians and scientists.

From early Greek philosophy, there has always been a great divide in thought. Humanity is faced with a choice between two metaphysics, two differing views of how the world works from the top down and, hence, of the ultimate means for the selection of moral codes. The first view holds that morality is transcendental in origin and exists both within and apart from the human species. This doctrine has been refined within the Church by the conception of natural law, which is the reading of the eternal law in God's mind: People reason out God's intent through a reflection of human nature, obedient to the principle that, as Aquinas expressed it, "man has a natural inclination to know the truth about God and to live in society." The opposing view is that morality is entirely a human phenomenon. In the modern, evolutionary version of this materialist philosophy, its precepts represent the upwelling of deep impulses that are encoded in our genes and find expression within the setting of particular cultures. They have nothing directly to do with divine guidance, at least not in the manner conceived by traditional religions.

I may be wrong (and, in any case, do not speak for all scientists), but I believe that the correct metaphysic is the materialist one. It works in the following way. Our profound impulses are rooted in a genetic heritage common to the entire species. They arose by evolution through natural selection over a period of tens or hundreds of thousands of years. These propensities provide survival for individuals and for the social groups on which personal survival depends. They are transmuted through rational processes and the formation of culture into specific moral codes that are integrated into religion and the sacralized memories of revolutions, conquests, and other historical events by which cultures secured their survival. Although variations in the final codes are inevitable, different societies share a great deal in their perception of right and wrong. By making the search for these similarities part of the scientific enterprise, and by taking religious behavior very seriously as a key part of genetically evolved human nature, a tighter consensus on ethical behavior might be reached.

Let me interpose, at this point, a very brief account of evolution by natural selection. Genetic variation among individuals in a population of the same species, say a population of human beings, arises by mutations, which are random changes in the chemical composition and relative positions of genes. Of the thousands of mutations that typically occur throughout a population in each generation, all but a minute fraction are either neutral in effect or deleterious to some degree. They include, for example, the altered genes that cause hemophilia and Tay-Sachs disease and the abnormally duplicated chomosomes responsible for Down's syndrome. When a new mutant (or novel combination of rare preexisting genes) happens to be superior to the ordinary "normal" genes, it tends to spread through the population over a period of many generations and, hence, become by definition the new genetic norm. If human beings were to move into a new environment that somehow gave hemophiliacs a survival and reproductive advantage over nonhemophiliacs, then, in time, hemophilia would predominate in the population and be regarded as the norm.

Two features of evolution by natural selection conspire to give it extraordinary creative potential. The first is the driving power of mutations. All populations are subject to a continuous rain of new genetic types that test the old. The second feature is the ability of natural selection to create immensely complicated new structures and physiological processes, including new patterns of behavior, with no blueprint and no force behind them other than the selection process itself. This is a key point missed by creationists and other critics of evolutionary theory, who often argue that the probability of assembling an eye or a hand (or life itself) by genetic mutations is infinitesimally small—in effect, impossible. But, the following thought experiment shows that the opposite is true. Suppose that a new trait emerges if two new gene forms (mutations), which I will call A and B, occur simultaneously. The chance of A occurring is one in a million, and the chance of B occurring is also one in a million. Then, the chance of both A and B occurring simultaneously as mutants is one in a trillion, a near impossibility—as the critics intuited. However, natural selection subverts this process. If A has even a slight advantage by itself alone, it will become the dominant gene at its position. Now, the chance of AB appearing is one in a million. In even moderately sized species of plants and animals (which often contain more than a million individuals), the change-over to AB is a virtual certainty.

This very simple picture of evolution at the level of the gene has altered our conception of both the nature of life and humanity's place in nature. Before Darwin, it was customary to use the great complexity of living organisms per se as proof of the existence of God. The most famous expositor of this "argument from design" was the Reverend William Paley, who in 1802 introduced the watchmaker analogy: The existence of a watch implies the existence of a watchmaker. In other words, great effects imply great causes. Common sense would seem to dictate the truth of this deduction, but common sense, as Einstein once noted, is only our accumulated experience up to the age of eighteen. Common sense tells us that one-ton satellites cannot hang suspended two hundred miles above a point on the earth's surface, but they do.

We have arrived at the conception of the one book of creation to which I alluded earlier. Given the combination of mutation and natural selection, the biological equivalent of watches can be created without a watchmaker. But, did blind natural selection also lead to the human mind, including moral behavior and spirituality? That is the grandmother of questions in both biology and the humanities. Common sense would seem, at first, to dictate the answer to be no. But, I and many other scientists believe that the answer may be yes. Furthermore, it is possible by this means to explain the very meaning of human life.

The key proposition based on evolutionary biology is the following: Everything human, including the mind and culture, has a material basis and originated during the evolution of the human genetic constitution and its interaction with the environment. To say this much is not to deny the great creative power of culture or to minimize the fact that most causes of human thought and behavior are still poorly understood. The important point is that modern biology already can account for many of the unique properties of our species. Research on that subject is accelerating quickly enough to lend plausibility to the proposition that more complex forms of social behavior, including religious belief and moral reasoning, will eventually be understood to their foundations.

A case in point, useful for its simplicity and tractability, is the avoidance of brother-sister incest. In order to avoid misunderstanding, let me define *incest* as strong sexual bonding among close biological relatives that includes intercourse of the kind generally associated with cohabitation and procreation and excludes transient forms of adolescent experimentation. Incest taboos are very nearly universal and a cultural norm. The avoidance of brother-sister incest originates in what psychologists have called "prepared learning." This means that people are innately prone to learn one alternative as opposed to another. They pick it up more readily, they enjoy it more, or both. The avoidance of sibling incest comes from the "potty rule" in mental development: Individuals reared in close domestic proximity during the first six years of life (they share the same potty) are automatically inhibited from strong sexual attraction and bonding when they reach sexual maturity. The rule works even when the children reared together are biologically unrelated and later encouraged to marry and have children, as in the Israeli kibbutzim and traditional minor marriages of prerevolutionary China. Those affected are usually quite unable to offer a rational explanation of why they have no attraction. Some unconscious process ticked over in the brain and the urge, they explain, never came.

The inhibitory rule is an example not only of prepared learning but also of "proximate causation" as it is understood by evolutionary biologists. This means that the learning channels a response of importance to the survival or reproduction of the organism. Proximate causes are put into place by the assembly of genes through the process of natural selection. The ultimate causation—in other words, the particular selection regime that enabled certain genes to predominate in the first place—is the well-documented effect of inbreeding depression. When mating occurs between brother and sister, father and daughter, or mother and son, the probability of matching debilitating genes in both homologous chromosomes of the offspring is greatly increased. The end result is a rise in abortion, physical defects, and genetic disease. Hence, genes prescribing a biological propensity to avoid incest will be favored over those that do not. Most animal and plant species display proximate devices of one kind or another, and it does indeed protect them from inbreeding depression. In some, the response is rigidly determined. In others, especially the brighter mammals, it is based on prepared learning. Interestingly enough, the human proximate form is nearly identical to that of the chimpanzee, the species to which we are most closely related genetically.

It is exquisitely human to semanticize innate tendencies. In many societies, incest avoidance is underwritten by symbolically transmitted taboos, myths, and laws. These, not the emotions and programs of prepared learning, are the values we perceive by direct, casual observation. They are easily transmitted from one person to the next, and they are the behaviors most readily studied by scholars. But, the phenomenon of greatest interest is the etiology of the moral behavior: The chain of events leading from ultimate cause in natural selection to proximate cause in prepared learning to reification and legitimation in culture. If the terminal cultural form were somehow to be stripped away by a collective loss of memory, people would still avoid sibling incest. Given enough time, they would most likely invent religious and ethical rationalizations to justify their feelings about the wrongness of incest.

Crude genetic determinism has no part in this process. The existence of the three-step etiology in mental development, genes to learning rules to culture, in no way contradicts free will. Individual choice persists, even when learning is strongly prepared by heredity. If some future society decides to encourage brother-sister incest, for whatever bizarre and unlikely reason, it now has the knowledge to do so efficiently. The possibility, however, is vanishingly remote, because the same knowledge tells us that incest avoidance is programmed as a powerful rule and protects families from genetic damage. We are likely to agree still more firmly than before that the avoidance is a part of human nature to be fostered. In short, incest avoidance is and will continue to be one of our common values.

It will immediately occur to you that incest avoidance might be no more than a special case in the evolution of social behavior. A vast difference separates this relatively simple phenomenon from economic cycles, religious rites, and presidential elections. Might such particularities fall within a wholly different domain of explanation and require a different metaethic? Perhaps, but I don't think so.

The evidence favoring the evolutionary approach to moral reasoning is as follows. By mid-1985, no fewer than 3,577 human genes had been identified, of which about six hundred had been placed on one or another of the twenty-three pairs of chromosomes. This is a respectable fraction of the entire human complement. New techniques for separation and identification make it possible to map most of the genes and specify all of the DNA sequences—perhaps, by early in the next century. Hundreds of the genes already known alter behavior in one way or another. In most cases, the effect is crude or indirect. But a few change behavior in a precise manner, as for example those modulating expression, reading ability, and performance on spatial tests. Twin and adoption studies have implicated other genes—as yet unmapped and probably working in complex multiples—in schizophrenia, propensity toward homosexuality, performance on tests measuring empathy, and a wide range of personality traits from introversion-extroversion to athleticism and proneness to alcoholism. Moreover, prepared learning and biases in perception have been discovered in virtually every category of behavior thus far studied. In their seminal book *The Biology of Religion,* Vernon Reynolds and Ralph Tanner showed that survival and genetic reproduction can be favored by the traditional practices of religion, including evangelism, marriage rites, and even celibacy and asceticism, the latter through their positive effects on group cohesion and welfare.

But, to come quickly to the point that most troubles critics of evolutionary ethics, it does not follow that the genetic programs of cognition and prepared learning are automatically beneficial, even in a crude Darwinian sense. Behaviors such as xenophobia and territorial expansion may have been very adaptive in the earlier, formative stages of human evolutionary history, but they are destructive now, even for those who practice them. Although the cultural *ought* is more tightly linked to the genetic *is* than philosophers have traditionally conceded, the two do not automatically translate one into the other. A workable moral code can be obtained not just by understanding the foundations of human nature but by the wise choice of those constraints needed to keep us alive and free in a rapidly changing cultural environment that renders some of our propensities maladaptive.

Let me illustrate this approach to moral reasoning by taking an example that has proved troublesome to the Church. In *Humanae Vitae,* Pope Paul VI used the best interpretation concerning human nature available to him to proscribe artificial birth control and to protect the family. He said, in effect, that you should not prevent conception when having sex because that is what sex is for and, as such, reflects the will of God: "To use this divine gift destroying, even if only partially, its meaning and its purpose is to contradict the nature both of man and of woman and of their most intimate relationship, and, therefore, it is to contradict also the plan of God and his will."

I believe that there is a way out of the impasse that this strict argument from natural law has created. All that we have learned of biology in recent years suggests that the perception of human nature expressed by Pope Paul VI was only half true. A second major function of sexual intercourse, one evolved over vast periods of time, is the bonding of couples in a manner that enhances the long-term care of children. Only a minute fraction of sexual acts can result in conception, but virtually all can tighten the conjugal bonds. Many circumstances can be imagined, and in fact exist, in which family planning by artificial birth control leads to an improvement of the bonding function while promoting the rearing of healthy, secure children.

If this more recent and better substantiated view of human sexuality is accepted, a revision of *Humanae Vitae* could easily be written that accomplishes the main purpose of Pope Paul VI and the modern Church, permits artificial birth control, and, in fact, serves as a model of the utilization of scientific findings by religious thinkers.

I am now going to close with a truly radical suggestion: The choice among the foundations of moral reasoning is not likely to remain arbitrary. Metaethics can be tested empirically. One system of ethics and, hence, one kind of religion is not as good as another. Not only are some less workable, they are, in the most profound sense, less human.

The corollary is that people can be educated readily only to a narrow range of ethical precepts. This leaves a choice between evolutionary ethics and transcendentalism. The idea of a genetic origin of moral codes can be further tested by a continuance of biological studies of complex human behavior, including religious thought itself. To the extent that the sensory and nervous systems appear to have evolved by natural selection or some other purely natural process, the evolutionary interpretation will be supported. To the extent that they do not appear to have evolved in this manner, or to the extent that complex human behavior cannot be linked to a physical basis in the sensory and nervous systems, the evolutionary explanation will have to be abandoned and a transcendental explanation sought.

Which position—scientific materialism or religious transcendentalism—proves correct will eventually make a very great difference in how humanity views itself and plans its future. But, for the years immediately ahead, this distinction makes little difference if the following overriding fact is realized: Human nature is, at the very least, far more a product of self-contained evolution than ordinarily conceded by philosophers and theologians. On the other hand, religious thought is far richer and more subtle than present-day science can explain—and too important to abandon. Meanwhile, the areas of common concern are vast, and the two enterprises can converge in most of the areas of practical moral reasoning at the same time that their practitioners disagree about the ultimate causes of human nature.

What, then, is the best relation between religion and science toward which we might aim? I would say an uneasy but fruitful alliance. The role of religion is to codify and put into enduring poetic form the highest moral values of a society consistent with empirical knowledge and to lead in moral reasoning. The role of science is to test remorselessly every conclusion about human nature and to search for the bedrock of ethics—by which I mean the material basis of natural law. Science faces in religion its most interesting challenge, while religion will find in science the necessary means to meet the modern age.

# Epilogue

That concludes my talk to the Catholic bishops. Now an additional thought: If I'm right about the deep biological origins of religion, no amount of debunking and confrontation will significantly alter the colossi of organized religion. They won't collapse like the Soviet empire, with the leaders declaring that they have seen the light; for billions of religionists and their leaders there will be no reverse epiphany; the sacred texts won't crumble like the Berlin Wall. But the great organized religions *can* be humanized. They can evolve, through friendly truth and reason as opposed to hostile truth and reason, laid forth in a conciliatory style. Superstition will rarely or never be repudiated, it will simply be abandoned during the gradual humanization of organized religion. It's been said that the great problems of history are never solved, they are merely forgotten. And so it will be during the Darwinian sifting of ideas, hastened by open communication, total honesty, and warmth of spirit, in the best tradition of humanism. For both sides, let the chips fall where they may.

# 11

# Religious Dogmatism and Materialist Reductionism versus Humanism

## Yves Galifret

There is much evidence to suggest that we have to find new ways to surmount the crisis that confronts our civilization at the end of the second millennium.

Our world certainly needs a new Renaissance; and yet, there is a paradox here. While technical, economic, cultural, political, and philosophical conditions have changed considerably since the fifteenth and sixteenth centuries, one element has remained constant: the Catholic Church. It would be more correct to say that, today, the Church is trying to become the Church it used to be when it fought Renaissance humanists and the Reformation.

Renaissance and Reformation should not be confused. It is known, for instance, that Francis I, king of France from 1515 to 1547, was a friend of the humanists (in order to compete with the sterile scholastics of the Sorbonne, he founded the Collège de France in 1530). However, this same Francis I opposed the Reformation. Many humanists, such as those who published Lucretius's works as early as 1473, in fact went much further in the way of free thought than the Protestants. On some occasions, the Protestants were the persecutors of the humanists (the humanist physician Michel Servet was sentenced by Protestant leader John Calvin to be burned). Today, descendants of these Protestants include the Fundamentalists, who are so active in the United States and who are at the forefront of antihumanism. Nevertheless, for the papacy of the Renaissance, humanists and Protestants were both objects of disapproval. It can be said that the Counter-Reformation and the Council of Trent were directed equally against the Protestants and the free minds of the Renaissance.

Historical parallels can be dangerous, and some famous aphorisms exist which encourage prudence. Conversely, people are often blamed for having no memory and being unable to learn from the past to avoid the traps of the present. In this case, the past is particularly eloquent; Counter-Reformation politics can indeed be useful to help understand the strategy of contemporary Rome in its politics of reconquest.

In the sixteenth century, the success of free thinking and the expansion of Protestantism in Europe placed the Church in a very critical situation. The authority of the pope was seriously threatened, and Rome decided to react by using all the means at its disposal. It first opted for force and terror: in 1542, Pope Paul III (Alexander Farnese) reconstituted the Inquisition under the authority of Cardinal Carafa, the future Paul IV. This was a return to medieval methods, tortures, and massacres (in Rome, the French massacre of Saint-Bartholomew was celebrated joyously in 1572 with a sumptuous Te Deum mass). In 1543, intellectual constraints and control of thinking were added to the physical constraints, when Paul III decided to create an *Index librorum prohibitorum,* a catalogue of forbidden books and publications. Paul IV established the first list in 1559, and the last one appeared in 1948. The practice was suppressed by the last Council.

Fortunately, today's pope, John-Paul II, is not likely to reconstitute the Inquisition. Nevertheless, it is interesting to see how the Church deals with modern laws. In a symposium organized jointly by the Parisian Bar Association and the archbishopric, entitled "Law, Liberty and Faith," the following incident occurred and was described in the French newspaper *Le Monde* in the following terms: "Without recalling the abuses and transgressions of the law committed throughout history in the name of religion, such as the inquisitorial terrors, one member of the audience nevertheless made a plea for 'the rights of the agnostic and of the free thinker to quietude.' " This is a perfectly justified claim, given the tremendous, indiscreet, and constant pressure exerted by religion on society and its institutions.

This intervention followed a talk by the Archbishop of Paris, who tried to show the dangers of substituting human reason for God as the foundation of law and of considering religious freedom simply as a part of freedom of opinion. According to *Le Monde,* the cardinal replied that affirmation of the rights of believers should not frighten non-believers. On the contrary, "it shows them to be the very source of a certitude and of an absolute which is in the heart of man." Behavior today is very different from sixteenth-century behavior, but Catholicism is still characterized by an arrogant certainty that it is "right," that everyone else is "wrong," and that it has a duty to proclaim this situation. The *Le Monde* article ends by pointing out that the cardinal failed to convince the whole of the audience that "showing" does not mean "imposing."

Nowadays the re-establishment of the Inquisition is unimaginable but this is far from being the case for the *Index.* An "instruction" on the press and publishing appeared recently, in which the Vatican urges bishops to prohibit—of what right and by what means?—the sale and exhibition of "publications related to religious or moral questions which have not received approval from the ecclesiastical authorities." The same text goes on to say that, henceforth, clerics and religious figures must obtain permission before publishing articles in "newspapers, opuscules or magazines which usually and overtly attack the Catholic religion and morality." This lumping together of religion and morality is a reflection of the old formula used, for example, when religion was fighting the "libertines."

The parallel between the contemporary Catholic Church and the Catholic Church of the Counter-Reformation does not end here. The Council of Trent (1545–1563), convoked by Paul III, approved decisions concerning the Inquisition and the *Index,* and adopted a series of measures designed first to bring back order to a Church which was in a very poor state, dissolute, and simoniacal, and second, to reinforce the fight against impiety and Protestantism. A Catholic historian, Jean Delumeau, professor at the Collège de France, commented that the Council of Trent was "a refusal to hold any dialogue with other Christians considered as heretics, and an unrestricted reinforcement of anti-Protestant positions."

It is interesting to compare the attitude of the Council of Trent with the attitude of Rome today as set forth in the "Letter on universality of the Church and primacy of the Pope." This letter was sent to the bishops, and sets out very clearly what is required of Protestant, Anglican, and orthodox churches: submission pure and simple; in other words, conversion. Cardinal Ratzinger states that "the primacy of Rome's bishop [the pope] and the episcopal college are elements of the universal Church and not some product of the peculiarities of churches." The pope's power must be accepted by all Christians as "the highest, absolute and universal" power, "present inside all churches." It could be said that, symbolically, Rome is "breaking in" and settling in the heart of all non-Catholic Christian churches. In ecumenical circles, this letter from Rome is bound to cause disappointment, since it cuts short speculation around the formulae of "federation" or "community of sister churches." It is a manifestation of the present pope's implacable and somewhat frightening desire for hegemony.

To top off the process, a universal catechism would be required, such as the *Catechism for the Priests,* written by the Council of Trent and published by Pope Pius V.* Here, reality outstrips fiction. This catechism actually exists, it is in the process of being printed and is due to appear shortly.† It is said to summarize, "the contents of the Catholic faith and morals," and according to some, the decision made in 1985 to write a universal catechism was interpreted as a castigation

---

*The Catholics have made a saint of this Dominican monk who was an inquisitor in Lombardy and then Great Inquisitor before becoming pope. He was the leader of the Holy Christian League which, by helping the Venetians, gained victory in Lepante (1571) over the Turks. This victory of Christians over Moslems was celebrated magnificently all over Christendom.

†It is not the subject of this paper to comment, but it is worth noting the approval of the death penalty in certain cases and the surrealist turn of the chapter concerning chastity.

of the various local catechisms, considered insufficiently rigorous and constrictive (in particular, the French episcopate was reprehended in Rome in 1983). The catechism is also interpreted as proof of the influence of traditionalist circles which, since the last Council, have conducted a campaign for a single, traditional catechism. In this way, the pope probably hopes to be able to check local particularities and ensure that everyone, everywhere, thinks as he would like them to do.

The role played by the Jesuits in the Catholic reconquest of the sixteenth century is a further illustration of the parallel between the popes of the Counter-Reformation and John Paul II. The Society of Jesus was organized and formally approved by Paul III in 1540. The initial objective for Ignatius of Loyola and his companions was to go to the "Holy Land" in order to convert the Infidels, but in fact Europe became their field of activity, since they turned their attention to the fight against the Reformation.

Their method was cautious, slow and yet efficient; they became teachers. They successfully took charge of the education of the sons of the upper classes and thus gained the allegiance of the leaders of the future. The period from 1551 to 1578 is generally accepted as the period during which the Jesuit colleges were founded in various countries—Austria, the Netherlands, Bavaria, Rhenish countries, Poland, the German Empire, etc.—and also the period which marks the end (in some cases definitive) of the growth of Protestantism and the beginning of the restoration of Catholic influence.

Today, looking back over this period, but relying only moderately on the Jesuits, John Paul II probably considers the Opus Dei as his potential semi-secret army dedicated to the carrying out of his project. This may explain the privileges accorded to this order, such as personal prelacy since 1982 and the unusually rapid beatification of the Opus founder, Escriva de Balaguer.

De Balaguer founded the Opus Dei in 1928, but the order began to make inroads only after the victory of Franco over the Spanish Republic. It exists today in many countries, and in particular seems to exert an influence in the spheres of politics, high finance, and industry. However, adapting to modern conditions for the sharing of real power is no substitute for the strategy initiated in the sixteenth century, when education of the younger generations was used as a means to spiritual and even temporal reconquest. It is merely an extension of that strategy. This must not be forgotten, particularly by Europeans. One of the pope's projects is to rebuild a Christian Europe, the Europe of the Counter-Reformation era, the Europe of the Council of Trent.

Lack of space prevents a long discussion of the problem related to Islamic fundamentalism. In this case there is no need to plan a return to the Middle Ages, these religions having not been confronted either by a Renaissance, a Reformation, or an Enlightment. The spreading of other sects is also a matter for concern especially in former Communist countries.

In the face of these mystical movements and growing irrationalism, it is our duty to propose a philosophy that avoids oversimplification and naiveté. This is a condition for credibility and effectiveness. As philosopher and epistemologist Georges Canguilhem[1] wrote, when describing an out-of-date rationalism:

> It was frequently a commitment of reason against religion or against the traditionalist established order rather than a commitment for the rationality of reason against its own tradition. This sort of commitment was based on an impassive reason certain to find and to recognize itself in the continuous progress of the science by which it was educated.

According to Canguilhem, true rationalism must be ". . . a commitment of reason against this sort of rationalism, a sort of scientific superstition, self-satisfied by the first successes of rationalization." Critical rationalism must play first against rationalism itself. Hellenist Jean-Pierre Vernant,[2] professor at the Collège de France and a member of the French Union Rationaliste, also condemns what he calls "a sort of metaphysics of reason" and exhorts caution in the face of "a certain conception of Reason, immutable, eternal, absolute," which still subsists in some rationalist circles.

This rough, elementary rationalism ignores the complexity of reality and refuses to progress as the scientific deciphering of this complexity progresses. Consequently, it cannot successfully oppose irrational spiritualism.

This ignorance of the complexity is well illustrated by the nineteenth-century materialist rationalist's answer to the puzzling mind-brain problem. In his book *Rapports du physique et du moral de l'homme* (1802), the physician Cabanis, member with Destutt de Tracy and Volney of the group of "Idéologues," writes that the brain "organically produces the secretion of thought," and later, Darwin[3] also speaks of thought as a "brain secretion." Moleschott[4] writes: "The thought is to the brain as the bile is to the liver and the urine to the kidneys." In the light of the cultural environment of their time, all these authors intend to say that mental activity is the result of brain activity and not due to the contribution of mysterious entities. Of course, they are right: it was necessary to react to the claims of a well-ensconced spiritualism. However, their formulae unfortunately provided a very poor image of reality. The term "secretion" which probably seemed significant then, now seems particularly inadequate, since it does not take account of the specific complexity of brain mechanisms. Moreover, the parallel between the brain, the liver, and the kidneys ignores the fundamental difference between these organs and the brain: the brain increases its functional capacities as it functions, such self-improvement being possible only through interaction with the environment. Although it is possible to envisage two people exchanging their kidneys or even their livers (the technique, clearly, is somewhat more complicated), it is difficult to imagine all the consequences resulting from an exchange of brains (in fact, technically this is impossible).

In short, the nineteenth-century materialists, in their fight against spiritualism, would have been more convincing if, rather than reducing the brain to the status of some organ, they had highlighted the fact that mental life—an exceptional product—must be derived from an exceptional machine. The enormous gap between the explanations offered by these authors and the infinite complexity of psychological life as experienced by every human being left the field open for the development of spiritualist psychology. This was the case in France with Henri Bergson[5] (for whom the mind is to the brain what the coat is to the hook on which it hangs), and in the United States with William James (his correspondence with Bergson is particularly interesting from this point of view).

Strict reductionism reveals its deficiencies more clearly when confronted with the eternal mind-body problem. Mario Bunge and Rubén Ardila[6] have written a book on the subject, *Philosophy of Psychology*, honoring modern rationalism. Their conclusion is worthy of attention; it enables the reader to appreciate, by comparison with the above quotations from the beginning of the nineteenth century, the importance of the progress achieved in the course of two centuries of rationalist thought.

> We have argued for the reductionist thesis that mental phenomena are biological processes, as well as for the emergentist thesis that mentation is a qualitative novelty emerging at certain points in the evolution of biopopulations and in the development of individuals of some animal species. Moreover, we have argued that the emergence of mental abilities can be explained, at least in principle, by identifying it with the organization or reorganization of neuronal systems (i.e., the change in connectivity) either spontaneously (without any external causes) or in response to changes occurring in other parts of the body or in the environment. We have thus combined ontological emergentism with moderate epistemological reductionism.
>
> However, in the case of behavior and mind, reduction is insufficient; it must be supplemented with a study of processes occurring in adjoining domains, sometimes higher-level ones. In particular, an adequate understanding of behavior and mentation in the case of gregarious animals calls for the cooperation of social science.

Paul Kurtz,[7] especially in his last brilliant and provocative book (1992), has proposed the term "coduction" "to describe how the behavioral and social sciences now seek to understand human behavior in its various dimensions."

The temptation of strict reductionism resisted by Mario Bunge and Rubén Ardila at the transition from the biological to the psychological level also exists at the transition from chemistry (today's molecular biology) to biology. Ernst Mayr,[8] the great evolutionary biologist from Harvard, writes: "Every biologist is fully aware of the fact that molecular biology has demonstrated decisively that all processes in living organisms can be explained in terms of physics or chemistry." Here, Mayr disavows any lingering attachment to vitalism, the first step in the rationalist process. The second step is a refusal of what he calls *explanatory reductionism:*

Where the argument begins is when one asks whether the knowledge of the basic constituents of an analysis would permit one to build up an understanding of the total system. We know that an inventory of all molecules in the liver is not sufficient to reconstruct a description of the entire liver . . . as one goes to the higher and higher levels of organization, new concepts emerge that are needed to understand the behavior at that level.[9]

The same problems occur in the epistemological relationship between chemistry and physics, and inside physics itself, between solid state physics and elementary particle physics. However, a more extreme kind of reductionism exists, labelled "philosophical reductionism" by Karl Popper. It is accepted without question among many scientists, as the following statement shows:

The supreme test of the physicist is to arrive at those universal elementary laws from which the cosmos can be built by pure deduction.

The author of this disputable statement is Albert Einstein. This statement was cited by another physicist and leader in the field of high-energy physics, Steven Weinberg, in *Scientific American* in 1974. He also wrote this:

One of man's enduring hopes has been to find a few simple laws that would explain why nature, with all its seeming complexity and variety, is the way it is. At the present moment the closest we can come to a unified view of nature is a description in terms of elementary particles and their mutual interactions.[10]

Ernst Mayr would say later[11] that this *Scientific American* article is "a horrible example of the way physicists think"!

A final example, from V. F. Weisskopf, also a physicist, is this:

Looking at the development of science in the Twentieth Century, one can distinguish two trends, which I will call "intensive" and "extensive" research, lacking a better terminology. In short: intensive research goes for the fundamental laws, extensive research goes for the explanation of phenomena in terms of known fundamental laws . . . solid state physics, plasma physics, and perhaps also biology are extensive. High energy physics and a good part of nuclear physics are intensive.[12]

Philip Anderson, a leading condensed-matter physicist, gave a pertinent refutation of Weisskopf's presentation of science:

> The ability to reduce everything to simple fundamental laws does not imply the ability to start from those laws and reconstruct the universe. . . . The constructionist hypothesis breaks down when confronted with the twin difficulties of scale and complexity. The behavior of large and complex aggregates of elementary particles, it turns out, is not to be understood in terms of a simple extrapolation of the properties of a few particles. Instead, at each level of complexity, entirely new properties appear, and the understanding of the new behaviors requires research which, I think, is as fundamental in its nature as any other. . . . At each stage, entirely new laws, concepts, and generalizations are necessary, requiring inspiration and creativity to just as great a degree as in the previous one. Psychology is not applied biology and biology applied chemistry.[13]

To avoid misinterpretation, it should be said that this does not exempt the phenomena described by psychology from obeying the laws of biology, and those described by biology from obeying the laws of chemistry. However, biology alone cannot explain psychology, and psychology alone cannot explain biology. As Anderson says "the whole becomes not only more than, but different from, the sum of its parts."

In fact, radical reductionism is absolutely untenable in its claim that the properties of a system are only the consequence of the properties of its constituents. Proof is provided by the radical reductionists when they say such things as "We are what our proteins are." Why stop at the protein level? Proteins are aminoacids, and aminoacids themselves are carbon, hydrogen, oxygen, nitrogen (and a touch of sulphur for the hair); logically, then, we must arrive at the inescapable conclusion that we are what carbon, oxygen, etc. are, which does not make sense.

More generally, if it were true that physics can explain chemistry, that chemistry can explain biology and so forth, right up to history, for instance, then we must wait for the moment when the Peloponnesian

War or the fall of the Soviet empire will become chapters of particle physics—which, of course, is nonsense.

If we are to elaborate a convincing form of rationalism as the basis of the new humanism needed by our time, such mistakes must be carefully avoided. We are faced with an enemy more dangerous than religious dogmatism here. That enemy is our own shortcomings.

## Notes

1. G. Canguilhem, Preface to G. Bachelard, *L'Engagement rationaliste* (Paris: PUF, 1972).

2. J. P. Vernant, "Raison d'hier et d'aujourd'hui," *Raison Présente* 55 (1980): 11–16.

3. Cited by P. Medawar in *Plato's Republic* (Oxford: Oxford University Press, 1982).

4. J. Moleschott, *Der Kreislauf des Lebens* (Mainz, 1852).

5. H. Bergson, *Matière et Mémoire;* avant-propos de la septième édition, in *Oeuvres* (éd. du centenaire) (Paris: PUF, 1911), pp. 161–68.

6. M. Bunge and R. Ardila, *Philosophy of Psychology* (Berlin-Heidelberg-New York: Springer Verlag, 1987).

7. P. Kurtz, *Decision and the Condition of Man* (Seattle: University of Washington Press, 1968, and New York: Dell Paperbacks, 1969); *The New Skepticism: Inquiry and Reliable Knowledge* (Buffalo, N.Y.: Prometheus Books, 1992).

8. E. Mayr, *The Growth of Biological Thought* (Cambridge: Harvard University Press, 1982).

9. E. Mayr and S. Weinberg, "The Limits of Reductionism," *Nature* 331 (1988): 475–76.

10. See also S. Weinberg, "Newtonianism, Reductionism and the Art of Testimony," *Nature* 330 (1987): 433–437.

11. E. Mayr, in *Evolution at a Crossroads,* ed. by D. J. Depew and B. H. Weber (Cambridge: MIT Press, 1985).

12. V. F. Weisskopf, *Brookhaven National Laboratory Publications, 888 T 360* (1965).

13. P. W. Anderson, "More Is Different," *Science* 177 (1972): 393–96.

# 12

# Neurobiological Bases of Beliefs

## José M. R. Delgado

The *Encyclopedia Britannica* (1943–1973) defines "belief" as "a mental attitude of acceptance or assert toward a proposition without the full intellectual knowledge required to guarantee its truth." Beliefs are based on interpretations, arrived at by intellectual judgment, or as indicated in the eighteenth century by David Hume, as a special sort of feeling. Beliefs become knowledge when the truth of a proposition has been objectively demonstrated.

The words "faith," "belief," and "conviction" are often considered to have similar meanings, although it may be preferable to differentiate them: Faith conveys a high degree of inner confidence, associated with supernatural revelation. Religious faith is experienced emotionally in the depth of personality, while religious beliefs entail the intellectual component of acceptance of dogma. Convictions may have stronger roots than beliefs and may persist in spite of their demonstrated falsehood. These semantic differentiations have no precise limits and may be considered rather arbitrary, but all are supported by neuronal mechanisms which can be investigated and known. In common language they are interchangeable.

Abstract thought is a typically human mental capability not shared with animals, requiring a high degree of neuronal complexity with the

179

power of abstraction, symbolism, and intelligence. Many animals, including dolphins, dogs, and monkeys, can be trained to obey orders and to perform astonishing tasks, but their mental capacity does not allow them to develop religious, political, or scientific beliefs typical of man.

Human beings function according to a large spectrum of beliefs, many of which are hidden in the subconscious beyond individual awareness, although they trigger many behavioral responses. To travel by air, for example, we assume that the plane and all its components have been well constructed, that the pilot has been well trained, and that the sophisticated technology at airports will work properly, although we lack proof, knowledge, and assurance that our expectations are valid. Similar comments may be expressed about television, surgical interventions, and even about banks accounts. Civilized life is very complex and is based on trust in many conditions about which we are unfamiliar.

## Beliefs, Brain, and Soul

The conflict between scientific discoveries and religious doctrines has many historical precedents and is even today a subject of deep concern for many people. In the seventeenth century, to support Copernicus's conception of the solar system with planets orbiting around the sun was heretical and condemned by the Church, which also rejected the mathematical and astronomical findings of Galileo. Doctrines of creation are contrary to Darwin's ideas about the evolution of species and natural selection. Consideration of the creation of life as a divine mystery was superseded when urea, an organic product, was synthesized in the laboratory, and when it was demonstrated that amino acids, the building blocks of proteins, can be obtained artificially. Modern neurobiology has been investigating the morphology and the physiology of the brain without detecting any signs of supernatural spirits, adopting in general materialistic positions.

The soul is defined in the *Encyclopedia Britannica* as "an immaterial principle or aspect that, with the body, constitutes the human person" and conceived as "the seat of the highest mental activities." The soul is closely associated with such terms as mind, spirit, or self. Life, breath, and movement traditionally emanate from the soul. Ancient Hebrews and Greeks regarded the soul as so indissolubly connected with the body that it ceased at death or persisted as only an attenuated, empty, shadowy entity. Plato and Plotinus viewed the soul as a pure and eternal element, pre-existing the body, which attained its full and perfect state only after separation from the body at death.

For Buddhists and Hindus, the soul is reincarnated in successive bodily existences, in a recurrent cycle from which release may only be obtained through moral, intellectual, or spiritual perfection. Talmudic Jews and Christians believe that souls will rejoin the bodies of the dead on a final glorious day when all human beings are resurrected.

In the eighteenth century, the German philosopher Immanuel Kant accepted the immortality of the soul as a practical support for morality and ethics. More recently the American philosopher and psychologist William James stated that the concept of soul was unnecessary for the understanding of personal identity and moral human values. This opinion is shared today by most scientists.

We may then ask: Who is right, the investigators, the philosophers, or the theologians? Since all of them have viewpoints related to their methodologies, we should interpret their statements as opinions of specialists and partial truths. The rigid attitude of some scientists is more surprising because they should be receptive to new information and to originality of interpretations.

Let us consider first the experimental fact that study of the brain does not reveal the existence of the soul or the spirit, providing no evidence to support the nearly universal belief that human personality has an important spiritual aspect.

Morphological research reveals the macroscopic, microscopic, and ultramicroscopic structure of the brain, with its static components of cells, pathways, and connections, including synaptic terminals and neuronal structures. These data show the intimate details of cerebral regions, but in general they have been obtained by observing dead tissues preserved in formalin or other substances, yielding valuable data concerning neuronal material carriers of signals, but not about functions of the living brain.

Neurophysiologists use a variety of technologies to investigate the activities of neurons, including electrical and chemical phenomena, action potentials, liberation of transmitters, and the complexity of metabolic phenomena that accompany behavioral manifestations, but they are not able to decipher the meaning of the signals circulating through the neuronal networks, because *the codes* used depend on *personal experience.* A classic example is the reading of a book: written words are codes representing ideas, objects, and situations. If we know the language used, for example, English, then we may decifer the codes circulating in our brains and understand the messages, but if the book is written in Chinese or in Arabic, and we have not learned these idioms, then we see only meaningless hieroglyphics. The action potentials evoked in the retina and optic nerve may be similar when looking at a specific word, even if we do not know the language, but the understanding of its meaning depends on our previous experience and learning stored in memory.

Electrical recordings of neuronal activity cannot reveal the symbolic content of information codes. The referential system, established by previous learning, is essential for comparison of new information with past experience, providing meaning and also emotional content to data reaching our sensory receptors.

*In summary:* The structure of material carriers and the patterning of electrical codes may be investigated by neurobiologists, while the meaning of symbols is a nonmaterial phenomenon related to personal experience and requiring multidisciplinary research.

Reinforcing these comments we may examine the functioning of magnetic tapes and video recorders. It is interesting to know the design, study the circuitry, analyze the electronic components, and register the electrical signals which circulate inside the instruments. These data, however, will not unveil the meaning of words or the sound of music. For this purpose we need the decodifying element of a person who can listen and understand. While the instruments may receive, store, and automatically reproduce any kind of idiom or music, the meaning of this information depends not on the electronic structure of the apparatus, but on the codification and subsequent decodification of the information received through the human sensory detectors.

The brain of the newborn child has a huge amount of neurons and pathways, with a capacity to learn any language and to receive a rich spectrum of sensory information, which will determine the organization of the personal referential system and contribute to the establishment of individual emotionality. Through these mechanisms, beliefs, ideologies, and ethical values will be inculcated in infantile minds, shaping neuronal structures and functions. These processes are automatic without the participation, knowledge, or consent of the individual because, at birth, the human brain is very immature and lacks the capacity to select or reject sensory information. Later on, when the "age of reason" is reached, the child begins to think about, modify, or even reject some behavioral reactions inculcated during infancy.

At present religious education is rather primitive, even in so-called civilized countries, where insufficient knowledge is provided about the great diversity of beliefs. In pre-Colombian times, the Aztecs had considerable religious uniformity, thinking that they were the descendants of the sun god and that it was necessary to offer good food to the sun (Tlaxcaltiliztli) to avoid its disappearance from the firmament. The survival of the sun depended on the providing of blood and live hearts and for these reasons human sacrifices were of paramount importance in religious rituals.

Beliefs in supernatural powers and beings have been accepted by most human societies from prehistoric times. If beliefs are so widely shared, even in the absence of intercultural contacts, they must have some common elements, perhaps related to biological factors and more specifically to neuronal structuring and functioning.

Human beliefs require the following elements: (a) a brain with genetic, morphological, and physiological normality to provide the material carrier with all its functions; (b) the coding of data transforming concepts into electrochemical activities; and (c) intracerebral decoding of messages in order to reach a conscious or subconscious understanding of meaning and to evoke perceptions, emotions, and behavior.

*In summary:* codes and meanings are *nonmaterial* entities which require material support, and both are the interrelated components of the psychophysical unity of each individual.

The wealth of beliefs, values, and skills acquired by each person represents a coded, nonmaterial content which can be transmitted to other beings, persisting even when the material carrier (the body) dies and disintegrates. In spite of its materialistic background, this reality supports the existence of nonmaterial phenomena, which may be considered as spiritual.

Acceptance or rejection of the theological significance of this spiritual entity and of a relation with deific powers depends on the neuronal structuring of our infantile referential system and on its subsequent cultural evolution.

Universality of belief in the supernatural is related to the neurobiological bases of learning. Human beings possess cerebral structures prepared for *acquisition* of language, manual skills, problem solving, and a great variety of functions, including abstract thought and the capacity to believe.

Supernatural beliefs have existed throughout recorded history, in primitive and civilized nations, in Nordic as well in tropical peoples, in minds with limited culture and in Nobel Prize winners. Different aspects of nature have been objects of adoration, including the sun, rivers, mountains, and forests. Many deities have been venerated, and most religions have similar laws, rewards, punishments, and liturgies based on the idea of a Supreme Being who is the immortal, infinite, wise creator of man and the universe. Many scientists who hold no personal belief in the supernatural do recognize the possibility of extrasensory perceptions, telepathy, psychic energy, and other unexplained phenomena.

Tentative explanations may be classified as follows:

a. The existence of supernatural entities, with or without religious implications, influences the world in some unknown and mysterious ways.

b. Supernatural powers are only a cultural misconception similar to other historical errors, such as supposing that the earth was flat and the center of the universe.

c. Supernatural beings are necessary to explain the existence of matter, energy, life, and the whole cosmos.

d. Development of neuronal functions requires sensory inputs with information originating in the environment; and development of individual frames of reference constitutes a major part of human personality.

Children must learn everything from walking and talking to complex behavior patterns. Receiving and expressing affection and also caring about others, present or absent, are very important elements for the structuring of the brain.

Lack of use leads to anatomical and functional atrophy. In children born blind because of congenital cataracts, visual neurons located at the occipital lobes remain underdeveloped or in the newborn stage. Something similar may occur with cerebral structures responsible for abstract thought and for the acquisition of beliefs. Lack of exposure to religion could result in closing of emotional experiences causing atrophy and neuronal rigidity manifested in the systems of beliefs in the adulthood.

## Education of Neurons

The fecundated ovum is a single cell with more than 100,000 genetic elements, constituting the programs that direct the formation of each human being. Speed of cellular multiplication is so astonishing that every minute 250,000 neurons may be created in the neocortex, leading to the formation of thousands of millions of brain cells.

The ovums's genetic wealth consists of programs for future development, which are possibilities but not realities, and initially it lacks heart and brain and therefore has no beliefs. Anthropologists, psychologists, and sociologists agree that both elements, genes and information, are totally necessary for cerebral maturation, brain development, and appearance of beliefs. Neurons can be educated by suitable training, especially during early childhood, and this capability should be known by educators and be used when teaching plans are developed. Support for this statement may be summarized as follows:

a. At birth, the human brain is undeveloped, has high plasticity, and is in the process of full neuronal growth. This fact and the lack of experience explain the poor motor coordination of the newborn, the impossibility of walking, the absence of language and of the many other human characteristics. At this time there is a total dependence on maternal care.

b. The "error of potentiality" is not acceptable. This error consists in considering that initially nonexisting functions will appear automatically in the future, whereas their development requires elements from the outside such as food and sensory stimuli.

c. Genetic heredity is not sufficient for the structuring of the baby. Information is essential for neuronal development.

d. Personal experience influences neuronal anatomy and physiology, establishing material traces within the individual system of reference.

e. The referential system, probably located in the limbic system, is necessary for the decodification and understanding of the information circulating inside of the brain.

f. The brain cannot be in direct contact with external reality. Information must be transduced at the sensory receptors as codes of electrical and chemical signals, which are the only ones able to circulate through neuronal pathways. These codes are the symbolic representation of reality.

g. The codes of sensory information leave material traces in the structure of the neurons, influencing their functions.

h. Education is necessary for the neuronal development of the brain.

i. War and peace are not programmed by genetic determination; they depend on intellectual and emotional elements provided by culture and learned by personal experience.

j. The brain has specific areas for the perception of personal pleasure. These areas may be activated more effectively by spiritual stimulation than by material satisfaction of primitive instincts.

Based on the above stated facts adults have a great responsibility in the structuring of infantile minds. Beliefs are not invented by each individual; they are a cultural heritage imparted by family, friends, and teachers. Intelligent education should choose suitable values for a proper neuronal education directed toward personal satisfaction compatible with social improvement.

# Note

Additional information may be found in the following publications:

Bloom, F. E., and A. Lazerson. *Brain, Mind, and Behavior.* New York: W. H. Freeman, 1988.

Changeux, J. P. *L'homme neuronal.* Arthème Fayard. Paris, 1983.

Churchland, P. S. *Neurophilosophy.* Cambridge, Mass.: MIT Press, 1992.

Delgado, J. M. R. *Physical Control of the Mind.* New York: Harper and Row, 1969.

———, *La Felicidad.* Madrid: Temas de Hoy, 1991.

Smith, A. *The Mind.* London: Penguin Books, 1984.

# 13

# Viruses of the Mind*

## Richard Dawkins

The haven all memes depend on reaching is the human mind, but a human mind is itself an artifact created when memes restructure a human brain in order to make it a better habitat for memes. The avenues for entry and departure are modified to suit local conditions, and strengthened by various artificial devices that enhance fidelity and prolixity of replication: native Chinese minds differ dramatically from native French minds, and literate minds differ from illiterate minds. What memes provide in return to the organisms in which they reside is an incalculable store of advantages—with some Trojan horses thrown in for good measure.

—Daniel Dennett

## Duplication-Fodder

A beautiful child close to me, six and the apple of her father's eye, believes that Thomas the Tank Engine really exists. She believes in Father

*Originally published in *Free Inquiry* (Summer 1993): 34–41. Reprinted with permission. This is an abridged version of an article in B. Dahlbom, ed., *Dennett and His Critics: Demystifying Mind* (Oxford: Blackwell, 1993).

Christmas, and when she grows up her ambition is to be a tooth fairy. She and her schoolfriends believe the solemn word of respected adults that tooth fairies and Father Christmas really exist. This little girl is of an age to believe whatever you tell her. If you tell her about witches changing princes into frogs she will believe you. If you tell her that bad children roast forever in hell she will have nightmares. I have just discovered that without her father's consent this sweet, trusting, gullible six-year-old is being sent, for weekly instruction, to a Roman Catholic nun. What chance has she?

A human child is shaped by evolution to soak up the culture of her people. Most obviously, she learns the essentials of their language in a matter of months. A large dictionary of words to speak, an encyclopedia of information to speak about, complicated syntactic and semantic rules to order the speaking, all are transferred from older brains into hers well before she reaches half her adult size. When you are preprogrammed to absorb useful information at a high rate, it is hard to shut out pernicious or damaging information at the same time. With so many mindbytes to be downloaded, so many mental codons to be duplicated, it is no wonder that child brains are gullible, open to almost any suggestion, vulnerable to subversion, easy prey to Moonies, scientologists, and nuns. Like immune-deficient patients, children are wide open to mental infections that adults might brush off without effort.

DNA, too, includes parasitic code. Cellular machinery is extremely good at copying DNA. Where DNA is concerned, it seems to have an eagerness to copy, like a child's eagerness to imitate the language of its parents. Concomitantly, DNA seems eager to be copied. The cell nucleus is a paradise for DNA, humming with sophisticated, fast, and accurate duplicating machinery.

Cellular machinery is so friendly toward DNA duplication that it is small wonder cells play host to DNA parasites—viruses, viroids, plasmids, and a riff-raff of other genetic fellow travelers. Parasitic DNA even gets itself spliced seamlessly into the chromosomes themselves. "Jumping genes" and stretches of "selfish DNA" cut or copy themselves out of chromosomes and paste themselves in elsewhere. Deadly oncogenes are almost impossible to distinguish from the legitimate genes between which they are spliced. In evolutionary time, there is probably a continual traffic from "straight" genes to "outlaw," and back again.[1] DNA is just DNA. The only thing that distinguishes viral DNA from host DNA is its expected method of passing into future generations. "Legitimate" host DNA is just DNA that aspires to pass into the next generation via the orthodox route of sperm or egg. "Outlaw" or parasitic DNA is just DNA that looks to a quicker, less cooperative route to the future, via a sneezed droplet or a smear of blood, rather than via a sperm or egg.

For data on a floppy disk, a computer is a humming paradise just as cell nuclei hum with eagerness to duplicate DNA. Computers and their associated disk and tape readers are designed with high fidelity in mind. As with DNA molecules, magnetized bytes don't literally "want" to be faithfully copied. Nevertheless, you can write a computer program that takes steps to duplicate itself. Not just duplicate itself within one computer but spread itself to other computers. Computers are so good at copying bytes, and so good at faithfully obeying the instructions contained in those bytes, that they are sitting ducks to self-replicating programs: wide open to subversion by software parasites. Any cynic familiar with the theory of selfish genes and memes would have known that modern personal computers, with their promiscuous traffic of floppy disks and e-mail links, were just asking for trouble. The only surprising thing about the current epidemic of computer viruses is that it has been so long in coming.

## Computer Viruses: A Model for an Informational Epidemiology

Computer viruses are pieces of code that graft themselves into existing, legitimate programs and subvert the normal actions of those programs. They may travel on exchanged floppy disks, or over networks. They are technically distinguished from "worms" which are whole programs in their own right, usually traveling over networks. Rather different are "Trojan horses," a third category of destructive programs, which are not themselves self-replicating but rely on humans to replicate them because of their pornographic or otherwise appealing content. Both viruses and worms are programs that actually say, in computer language, "Duplicate me." Both may do other things that make their presence felt and perhaps satisfy the hole-in-corner vanity of their authors. These side effects may be "humorous" (like the virus that makes the Macintosh's built-in loudspeaker enunciate the words "Don't panic," with predictably opposite effect); malicious (like the numerous IBM viruses that erase the hard disk after a sniggering screen-announcement of the impending disaster); political (the Spanish Telecom and Beijing viruses protest about telephone costs and massacred students respectively); or simply inadvertent (the programmer is incompetent to handle the low-level system calls required to write an effective virus or worm). The famous Internet Worm, which paralyzed much of the computer power of the United States on November 2, 1988, was not intended (very) maliciously but got out of control and, within twenty-four hours, had clogged around six thousand computer memories with exponentially multiplying copies of itself.

"Memes now spread around the world at the speed of light, and replicate at rates that make even fruit flies and yeast cells look glacial in comparison. They leap promiscuously from vehicle to vehicle, and from medium to medium, and are proving to be virtually unquarantinable."[2] Viruses aren't limited to electronic media such as disks and data lines. On its way from one computer to another, a virus may pass through printing ink, light rays in a human lens, optic nerve impulses, and finger muscle contractions. A computer fanciers' magazine that printed the text of a virus program for the interest of its readers has been widely condemned. Indeed, such is the appeal of the virus idea to a certain kind of puerile mentality (the masculine gender is used advisedly), that publication of any kind of "how to" information on designing virus programs is rightly seen as an irresponsible act.

I am not going to publish any virus code. But there are certain tricks of effective virus design that are sufficiently well-known, even obvious, that it will do no harm to mention them, as I need to do in order to develop my theme. They all stem from the virus's need to evade detection while it is spreading.

A virus that clones itself too prolifically within one computer will soon be detected because the symptoms of clogging will become too obvious to ignore. For this reason many virus programs check, before infecting a system, to make sure that they are not already on that system.

Incidentally, this opens the way for a defense against viruses that is analogous to immunization. In the days before a specific antivirus program was available, I myself responded to an early infection of my own hard disk by means of a crude "vaccination." Instead of deleting the virus that I had detected, I simply disabled its coded instructions, leaving the "shell" of the virus with its characteristic external "signature" intact. In theory subsequent members of the same virus species that arrived in my system should have recognized the signature of their own kind and refrained from trying to double-infect. I don't know whether this immunization really worked, but in those days it probably was worthwhile "gutting" a virus and leaving a shell like this, rather than simply removing it lock, stock, and barrel. Nowadays it is better to hand the problem over to one of the professionally written antivirus programs.

A virus that is too virulent will be rapidly detected and scotched. A virus that instantly and catastrophically sabotages every computer in which it finds itself will not find itself in many computers. It may have a most amusing effect on one computer: erase an entire doctoral thesis or something equally side-splitting—but it won't spread as an epidemic.

Some viruses, therefore, are designed to have an effect that is small enough to be difficult to detect, but which may nevertheless be extremely damaging. There is one type that, instead of erasing disk sectors wholesale, attacks only spreadsheets, making a few random changes in the (usually financial) quantities entered in the rows and columns. Other viruses evade detection by being triggered probabilistically, for example erasing only 1 in 16 of the hard disks infected. Yet other viruses employ the time-bomb principle. Most modern computers are "aware" of the date, and viruses have been triggered to manifest themselves all around the world, on a particular date such as Friday 13 or April Fool's Day. From the parasitic point of view, it doesn't matter how catastrophic the eventual attack is, provided the virus has had plenty of opportunity to spread first, a disturbing analogy to the Medawar/Williams theory of ageing; we are the victims of lethal and sublethal genes that mature only after we have had plenty of time to reproduce.[3] In defense, some large companies go so far as to set aside one "miner's canary" among their fleet of computers, and advance its internal calendar a week so that any time-bomb viruses will reveal themselves prematurely before the big day.

Again predictably, the epidemic of computer viruses has triggered an arms race. Antiviral software is doing a roaring trade. These antidote programs—"Interferon," "Vaccine," "Gatekeeper," and others—employ a diverse armory of tricks. Some are written with specific, known and named viruses in mind. Others intercept any attempt to meddle with sensitive systems areas of memory and warn the user.

The virus principle could in theory be used for nonmalicious, even beneficial purposes.[4] Looking into the future, it is not fanciful to imagine a time when viruses, both bad and good, have become so ubiquitous that we could speak of an ecological community of viruses and legitimate programs coexisting in the silicosphere. At present, software is advertised as, say, "Compatible with System 7." In the future, products may be advertised as "Compatible with all viruses registered in the 1988 World Virus Census, immune to all listed virulent viruses, takes full advantage of the facilities offered by the following benign viruses if present. . . ." Word-processing software, say, may hand over particular functions, such as word-counting and string-searches, to friendly viruses burrowing autonomously through the text.

Looking even further into the future, whole integrated software systems might grow, not by design, but by something like the growth of an ecological community such as a tropical rain forest. Gangs of mutually compatible viruses might grow up, in the same way as genomes can be regarded as gangs of mutually compatible genes.[5] Indeed, I have even suggested that our genomes should be regarded as gigantic colonies of viruses.[6] Genes cooperate with one another in genomes because natural selection has favored those genes that prosper in the presence of the other genes that happen to be common in the gene pool. Different gene pools may evolve toward different combinations of mutually compatible genes. I envisage a time when, in the same kind of way, computer viruses may evolve toward compatibility with other viruses, to form communities or gangs.

But then again, perhaps not! At any rate, I find the speculation more alarming than exciting.

At present, computer viruses don't strictly evolve. They are invented by human programmers and if they evolve they do so in the same weak sense as cars or airplanes evolve. Designers derive this year's car as a slight modification of last year's car, and they may, more or less consciously, continue a trend of the last few years—further flattening of the radiator grill or whatever it may be. Computer virus designers dream up ever more devious tricks for outwitting the programmers of antivirus software. But computer viruses don't—so far—mutate and evolve by true natural section. They may do so in the future. Whether they evolve by natural selection, or whether their evolution is steered by human designers, may not make much difference to their eventual performance. By either kind of evolution, we expect them to become better at concealment, and we expect them to become subtly compatible with other viruses that are at the same time prospering in the computer community.

DNA viruses and computer viruses spread for the same reason: an environment exists in which there is machinery well set up to duplicate and spread them around and to obey the instructions that the viruses embody. These two environments are, respectively, the environment of cellular physiology and the environment provided by a large community of computers and data-handling machinery. Are there are other environments like these, any other humming paradises of replication?

## The Infected Mind

I have already alluded to the programmed-in gullibility of a child, so useful for learning language and traditional wisdom, and so easily subverted by nuns, Moonies, and their ilk. More generally, we all exchange information with one another. We don't exactly plug floppy disks into slots in one another's skulls, but we exchange sentences, both through our ears and through our eyes. We notice each other's styles of moving and of dressing and are influenced. We take in advertising jingles, and are presumably persuaded by them, otherwise hardheaded businessmen would not spend so much money polluting the air with them.

Think about the two qualities that a virus, or any sort of parasitic replicator, demands of a friendly medium, the two qualities that make cellular machinery so friendly toward parasitic DNA, and that make computers so friendly toward computer viruses. These qualities are, first, a readiness to replicate information accurately, perhaps with some mistakes that are subsequently reproduced accurately; and, second, a readiness to obey instructions encoded in the information so replicated.

Cellular machinery and electronic computers excel in both these virus-friendly qualities. How do human brains match up? As faithful duplicators they are certainly less perfect than either cells or electronic computers. Nevertheless, they are still pretty good, perhaps about as faithful as an RNA virus, though not as good as DNA with all its elaborate proofreading measures against textual degradation. Evidence of the fidelity of brains, especially child brains, as data duplicators is provided by language itself. Shaw's Professor Higgins was able by ear alone to place Londoners in the street where they grew up. Fiction is not evidence for anything, but everyone knows that Higgins's fictional skill is only an exaggeration of something we can all do. Any American can tell Deep South from Midwest, New England from Hillbilly. Any New Yorker can tell Bronx from Brooklyn. Equivalent claims could be substantiated for any country. What this phenomenon means is that human brains are capable of pretty accurate copying (otherwise the accent of, say, Newcastle would not be stable enough to be recognized) but with some mistakes (otherwise pronunciation would not evolve, and all speakers of a language would inherit identically the same accents from their remote ancestors). Language evolves because it has the great stability and the slight changeability that are prerequisites for any evolving system.

The second requirement of a virus-friendly environment—that it should obey a program of coded instructions—is again only quantitatively less true for brains than for cells or computers. We sometimes obey orders from one another, but also we sometimes don't. Nevertheless, it is a telling fact that, the world over, the vast majority of children follow the religion of their parents rather than any of the other available religions. Instructions to genuflect, to bow toward Mecca, to nod one's head rhythmically toward the wall, to shake like a maniac, to "speak in tongues"—the list of such arbitrary and pointless motor patterns offered by religion alone is extensive—are obeyed, if not slavishly, at least with some reasonably high statistical probability.

Less portentously, and again especially prominent in children, the "craze" is a striking example of behavior that owes more to epidemiology than to rational choice. Yo-yos, hula hoops, and pogo sticks, with their associated behavior fixed patterns, sweep through schools, and more sporadically leap from school to school, in ways that differ from a measles epidemic in no serious particular. Ten years ago, you could have traveled thousands of miles through the United States and never seen a baseball cap turned back to front. Today the reverse baseball cap is ubiquitous. I do not know what the pattern of geographic spread of the reverse baseball cap was precisely, but epidemiology is certainly among the professions primarily qualified to study it. We don't have to get into arguments about "determinism"; we don't have to claim that children are compelled to imitate their fellows' hat fashions. It is enough that their hat-wearing behavior, as a matter of fact, is statistically affected by the hat-wearing behavior of their fellows.

Trivial though they are, crazes provide us with yet more circumstantial evidence that human minds, especially perhaps juvenile ones, have the qualities that we have singled out as desirable for an informational parasite. At the very least the mind is a plausible candidate for infection by something like a computer virus, even if it is not quite such a parasite's dream-environment as a cell nucleus or an electronic computer.

It is intriguing to wonder what it might be like, from the inside, if one's mind were the victim of a "virus." This might be a deliberately designed parasite, like a present-day computer virus. Or it might be an inadvertently mutated and unconsciously evolved parasite. Either way, especially if the evolved parasite was the memic descendant of a long line of successful ancestors, we are entitled to expect the typical "mind virus" to be pretty good at its job of getting itself successfully replicated.

Progressive evolution of more effective mind-parasites will have two aspects. New "mutants" (either random or designed by humans) that are better at spreading will become more numerous. And there will be a ganging up of ideas that flourish in one another's presence, ideas that mutually support one another just as genes do and as I have speculated computer viruses may do one day. We expect that replicators will go around together from brain to brain in mutually compatible gangs. These gangs will come to constitute a package, which may be sufficiently stable to deserve a collective name such as Roman Catholicism or Voodoo. It doesn't too much matter whether we analogize the whole package to a single virus, or each one of the component parts to a single virus. The analogy is not that precise anyway, just as the distinction between a computer virus and a computer worm is nothing to get worked up about. What matters is that minds are friendly environments to parasitic, self-replicating ideas or information, and that minds are typically massively infected.

Like computer viruses, successful mind viruses will tend to be hard for their victims to detect. If you are the victim of one, the chances are that you won't know it, and may even vigorously deny it. Accepting that a virus might be difficult to detect in your own mind, what telltale signs might you look out for? I shall answer by imagining how a medical textbook might describe the typical symptoms of a sufferer (arbitrarily assumed to be male).

1. The patient typically finds himself impelled by some deep, inner conviction that something is true, or right, or virtuous: a conviction that doesn't seem to owe anything to evidence or reason, but which, nevertheless, he feels as totally compelling and convincing. We doctors refer to such a belief as "faith."

2. Patients typically make a positive virtue of faith's being strong and unshakable, in spite of not being based upon evidence. Indeed, they may feel that the less evidence there is, the more virtuous the belief (see below).

This paradoxical idea that lack of evidence is a positive virtue where faith is concerned has something of the quality of a program that is self-sustaining, because it is self-referential.[7] Once the proposition is believed, it automatically undermines opposition to itself. The "lack of evidence is a virtue" idea would be an admirable sidekick, ganging up with faith itself in a clique of mutually supportive viral programs.

3. A related symptom, which a faith-sufferer may also present, is the conviction that "mystery," *per se,* is a good thing. It is not a virtue

to solve mysteries. Rather we should enjoy them, even revel in their insolubility.

Any impulse to solve mysteries could be seriously inimical to the spread of a mind virus. It would not, therefore be surprising if the idea that "mysteries are better not solved" was a favored member of a mutually supporting gang of viruses. Take the "Mystery of the Transubstantiation." It is easy and nonmysterious to believe that in some symbolic or metaphorical sense the eucharistic wine turns into the blood of Christ. The Roman Catholic doctrine of transubstantiation, however, claims far more. The "whole substance" of the wine is converted into the blood of Christ; the appearance of wine that remains is "merely accidental," "inhering in no substance."[8] Transubstantiation is colloquially taught as meaning that the wine "literally" turns into the blood of Christ. Whether in its obfuscatory Aristotelian or its franker colloquial form, the claim of transubstantiation can be made only if we do serious violence to the normal meanings of words like *substance* and *literally*. Redefining words is not a sin, but, if we use words like *whole substance* and *literally* for this case, what word are we going to use when we really and truly want to say that something did actually happen? As Anthony Kenny observed of his own puzzlement as a young seminarian, "For all I could tell, my typewriter might be Benjamin Disraeli transubstantiated. . . ."

Roman Catholics whose belief in infallible authority compels them to accept that wine becomes physically transformed into blood despite all appearances refer to the "mystery" of the transubstantiation. Calling it a mystery makes everything O.K. you see. At least, it works for a mind well prepared by background infection. Exactly the same trick is performed in the "mystery" of the Trinity. Mysteries are not meant to be solved, they are meant to strike awe. The "mystery is a virtue" idea comes to the aid of the Catholic, who would otherwise find intolerable the obligation to believe the obvious nonsense of the transubstantiation and the "three-in-one." Again, the belief that "mystery is a virtue" has a self-referential ring. As Hofstadter might put it, the very mysteriousness of the belief moves the believer to perpetuate the mystery.

An extreme symptom of "mystery is a virtue" infection is Tertullian's "Certum est quia impossibile est" (It is certain because it is impossible). That way madness lies. One is tempted to quote Lewis Carroll's White

Queen, who, in response to Alice's "One can't believe impossible things" retorted, "I daresay you haven't had much practice. . . . When I was your age, I always did it for half-an-hour a day. Why, sometimes I believed as many as six impossible things before breakfast." Or Douglas Adams's Electric Monk, a labor-saving device programmed to do your believing for you, which was capable of "believing things they'd have difficulty believing in Salt Lake City" and which, at the moment of being introduced to the reader, believed, contrary to all evidence, that everything in the world was a uniform shade of pink. But White Queens and Electric Monks become less funny when you realize that these virtuoso believers are indistinguishable from revered theologians in real life. "It is by all means to be believed, because it absurd" (Tertullian again). Sir Thomas Browne quotes Tertullian with approval, and goes further: "Methinks there be not impossibilities enough in religion for an active faith." And "I desire to exercise my faith in the difficultest point: for to credit ordinary and visible objects is not faith, but persuasion."[9]

I have the feeling that something more interesting is going on here than just plain insanity or surrealist nonsense, something akin to the admiration we feel when we watch a ten-ball juggler on a tightrope. It is as though the faithful gain prestige through managing to believe even more ridiculous things than their rivals succeed in believing. Are these people testing—exercising—their believing muscles, training themselves to believe impossible things so that they can take in their stride the merely improbable things that they are ordinarily called upon to believe?

While I was writing this, the *Guardian* (July 29, 1991) fortuitously carried a beautiful example. It came in an interview with a rabbi undertaking the bizarre task of vetting the kosher-purity of food products right back to the ultimate origins of their minutest ingredients. He was currently agonizing over whether to go all the way to China to scrutinize the menthol that goes into cough sweets. "Have you ever tried checking Chinese menthol. . . . It was extremely difficult, especially since the first letter we sent received the reply in best Chinese English, 'The product contains no kosher. . . .' China has only recently started opening up to kosher investigators. The methol should be OK, but you can never be absolutely sure unless you visit." These kosher investigators run a telephone hotline on which up-to-the-minute red-alerts of suspicion are recorded against chocolate bars and cod-liver oil. The rabbi sighs that the green-inspired trend away from artificial colors and flavors "makes life miserable in the kosher field because you have to follow all these things back." When the interviewer asks him why he bothers with this obviously pointless exercise, he makes it very clear that the point is precisely that there *is* no point:

> That most of the Kashrut laws are divine ordinances without reason given is 100 per cent the point. It is very very easy not to murder people. Very easy. It is a little bit harder not to steal because one is tempted occasionally. So that is no great proof that I believe in God or am fulfilling His will. But, if He tells me not to have a cup of coffee with milk in it with my mincemeat and peas at lunchtime, that is a test. The only reason I am doing that is because I have been told to do so. It is doing something difficult.

Helena Cronin has suggested to me that there may be an analogy here to Zahavi's handicap theory of sexual selection and the evolution of signals.[10] Long unfashionable, even ridiculed,[11] Zahavi's theory has been recently rehabilitated[12] and is now taken seriously by evolutionary biologists.[13] Zahavi suggests that peacocks, for instance, evolve their absurdly burdensome fans with their ridiculously conspicuous (to predators) colors, precisely because they are burdensome and dangerous, and therefore impressive to females. The peacock is, in effect, saying: "Look how fit and strong I must be, since I can afford to carry around this preposterous tail."

To avoid misunderstanding of the subjective language in which Zahavi likes to make his points, I should add that the biologist's convention of personifying the unconscious actions of natural selection is taken for granted here. Grafen has translated the argument into an orthodox Darwinian mathematical model, and it works. No claim is here being made about the intentionality or awareness of peacocks and peahens. They can be as sphexish or as intentional as you please.[14] Moreover, Zahavi's theory is general enough not to depend upon a Darwinian underpinning. A flower advertising its nectar to a "skeptical" bee could benefit from the Zahavi principle. But so could a human salesman seeking to impress a client.

The premise of Zahavi's idea is that natural selection will favor skepticism among females (or among recipients of advertising messages generally). The only way for a male (or any advertiser) to authenticate his boast of strength (quality, or whatever it is) is to prove that it is true by shouldering a truly costly handicap—a handicap *that only a genuinely strong* (high quality, etc.) male could bear. It may be called the "principle of costly authentication." And now to the point. Is it possible that some religious doctrines are favored not *in spite of* being ridiculous but precisely *because* they are ridiculous? Any wimp in religion could believe that bread *symbolically* represents the body of Christ, but it takes a real red-blooded Catholic to believe something as daft as the transubstantiation. If you can believe that you can believe anything, and (witness the story of Doubting Thomas) these people are trained to see that as a virtue.

Let us return to our list of symptoms that someone afflicted with the mental virus of faith, and its accompanying gang of secondary infections, may expect to experience.

4. The sufferer may find himself behaving intolerantly toward vectors of rival faiths, in extreme cases even killing them or advocating their deaths. He may be similarly violent in his disposition toward apostates (people who once held the faith but have renounced it); or toward heretics (people who espouse a different—often, perhaps significantly, only very slightly different—version of the faith). He may also feel hostile toward other modes of thought that are potentially inimical to his faith, such as the method of scientific reason that may function rather like a piece of antiviral software.

The threat to kill the distinguished novelist Salman Rushdie is only the latest in a long line of sad examples. On the very same day that I wrote this, the Japanese translator of *The Satanic Verses* was found murdered, a week after a near-fatal attack on the Italian translator of the same book. By the way, the apparently opposite symptom of "sympathy" for Muslim "hurt," voiced by by the Archbishop of Canterbury and other Christian leaders (verging, in the case of the Vatican, on outright criminal complicity) is, of course, a manifestation of the symptom we diagnosed earlier: the delusion that faith, however obnoxious its results, has to be respected simply because it *is* faith.

Murder is an extreme, of course. But there is an even more extreme symptom, and that is suicide in the militant service of a faith. Like a soldier ant programmed to sacrifice its life for germ-line copies of the genes that did the programming, a young Arab or Japanese is taught that to die in a holy war is the quickest way to heaven. Whether the leaders who exploit him really believe this does not diminish the brutal power that the "suicide mission virus" wields on behalf of the faith. Of course suicide, like murder, is a mixed blessing: would-be converts may be repelled, or may treat with contempt a faith that is perceived as insecure enough to need such tactics.

More obviously, if too many individuals sacrifice themselves the supply of believers could run low. This was true of a notorious example of faith-inspired suicide, though in this case it was not "kamikaze" death in battle. The Peoples' Temple sect went extinct when its leader, the Reverend Jim Jones, led the bulk of his followers from the United States ot the Promised Land of "Jonestown" in the Guyana jungle where he persuaded more than nine hundred of them, children first, to drink cyanide. The macabre affair was fully investigated by a team from the *San Francisco Chronicle.*[15]

> Jones, "the Father," had called his flock together and told them it was time to depart for heaven.
> "We're going to meet," he promised, "in another place."
> The words kept coming over the camp's loudspeakers.
> "There is great dignity in dying. It is a great demonstration for everyone to die."

Incidentally, it does not escape the trained mind of the alert sociobiologist that Jones, within his sect in earlier days "proclaimed himself the only person permitted to have sex" (presumably his partners were also permitted). A secretary would arrange for Jones's liaisons. She would call up and say, "Father hates to do this, but he has this tremendous urge and could you please. . . ?" His victims were not only female. One seventeen-year-old male follower, from the days when Jones's community was still in San Francisco, told how he was taken for dirty weekends to a hotel where Jones received a "minister's discount for Rev. Jim Jones and son." The same boy said:

> I was really in awe of him. He was more than a father. I would have killed my parents for him.

What is remarkable about the Reverend Jim Jones is not his own self-serving behavior but the almost superhuman gullibility of his followers. Given such prodigious credulity, can anyone doubt that human minds are ripe for malignant infection?

Admittedly, the Reverend Jones conned only a few thousand people. But his case is an extreme, the tip of the iceberg. The same eagerness to be conned by religious leaders is widespread. Most of us would have been prepared to bet that nobody could get away with going on television and saying, in all but so many words, "Send me your money, so that I can use it to persuade other suckers to send me their money too." Yet today, in every major conurbation in the United States, you can find at least one television evangelist channel entirely devoted to this transparent confidence trick. And they get away with it in sackfuls. Faced with suckerdom on this awesome scale, it is hard not to feel a grudging sympathy with the shiny-suited conmen. Until you realize that not all the suckers are rich, and that it is often widows' mites on which the evangelists are growing fat. I have even heard one of them explicitly invoking the principle that I now identify with Zahavi's principle of costly authentication. God really appreciates a donation, he said with passionate sincerity, only when that donation is so large that it hurts. Elderly paupers were literally wheeled on to testify how much happier they felt since they had made over their little all to the reverend whoever it was.

5. The patient may notice that the particular convictions that he holds, while having nothing to do with evidence, do seem to owe a great deal to epidemiology. Why, he may wonder, do I hold this set of convictions rather than that set? Is it because I surveyed all the world's faiths and chose the one whose claims seemed most convincing? Almost certainly not. If you have a faith, it is statistically overwhelmingly likely

that it is the same faith as your parents and grandparents had. No doubt soaring cathedrals, stirring music, moving stories, and parables help a bit. But by far the most important variable determining your religion is the accident of birth. The convictions that you so passionately believe would have been a completely different, and largely contradictory, set of convictions, if only you had happened to be born in a different place. Epidemiology, not evidence.

6. If the patient is one of the rare exceptions who follows a different religion from his parents, the explanation may still be epidemiological. To be sure, it is *possible* that he dispassionately surveyed the world's faiths and chose the most convincing one. But it is statistically more probable that he has been exposed to a particularly potent infective agent—a John Wesley, a Jim Jones, or a St. Paul. Here we are talking about horizontal transmission, as in measles. Before, the epidemiology was that of vertical transmission, as in Huntington's Chorea.

7. The internal sensations of the patient may be startlingly reminiscent of those more ordinarily associated with sexual love. This is an extremely potent force in the brain, and it is not surprising that some viruses have evolved to exploit it. St. Teresa of Avila's famously orgasmic vision is too notorious to need quoting again. More seriously, and on a less crudely sensual plane, the philosopher Anthony Kenny provides moving testimony to the pure delight that awaits those that manage to believe in the mystery of the transubstantiation. After describing his ordination as a Roman Catholic priest, empowered by laying on of hands to celebrate Mass, he goes on that he vividly recalls:

> . . . the exaltation of the first months during which I had the power to say Mass. Normally a slow and sluggish riser, I would leap early out of bed, fully awake and full of excitement at the thought of the momentous act I was privileged to perform. I rarely said the public Community Mass: most days I celebrated alone at a side altar with a junior member of the College to serve as acolyte and congregation. But that made no difference to the solemnity of the sacrifice or the validity of the consecration.
>
> It was touching the body of Christ, the closeness of the priest to Jesus, which most enthralled me. I would gaze on the Host after

the words of consecration, soft-eyed like a lover looking into the eyes of his beloved. . . . Those early days as a priest remain in my memory as days of fulfillment and tremulous happiness; something precious, and yet too fragile to last, like a romantic love-affair brought up short by the reality of an ill-assorted marriage.[16]

Dr. Kenny is affectingly believable that it felt to him, as a young priest, as though he was in love with the consecrated host. What a brilliantly successful virus! On the same page, incidentally, Kenny also shows us that the virus is transmitted contagiously—if not literally then at least in some sense—from the palm of the infecting bishop's hand through the top of the new priest's head: "If Catholic doctrine is true, every priest validly ordained derives his orders in an unbroken line of laying on of hands, through the bishop who ordains him, back to one of the twelve Apostles. . . . There must be centuries-long, recorded chains of layings on of hands. It surprises me that priests never seem to trouble to trace their spiritual ancestry in this way, finding out who ordained their bishop, and who ordained him, and so on to Julius II or Celestine V or Hildebrand, or Gregory the Great, perhaps."[17] It surprises me, too.

## Is Science a Virus?

No. Not unless all computer programs are viruses. Good, useful programs spread because people evaluate them, recommend them, and pass them on. Computer viruses spread solely because they embody the coded instructions: "Spread me." Scientific ideas, like all memes, are subject to a kind of natural selection, and this might look superficially viruslike. But the selective forces that scrutinize scientific ideas are not arbitrary or capricious. They are exacting, well-honed rules, and they do not favor pointless self-serving behavior. They favor all the virtues laid out in textbooks of standard methodology: testability, evidential support, precision, quantifiability, consistency, intersubjectivity, repeatability, universality, progressiveness, independence of cultural milieu, and so on. Faith spreads despite a total lack of every single one of these virtues.

You may find elements of epidemiology in the spread of scientific ideas, but it will be largely descriptive epidemiology. The rapid spread of a good idea through the scientific community may even look like a description of a measles epidemic. But when you examine the underlying reasons you find that they are good ones, satisfying the demanding standards of scientific method. In the history of the spread of faith you will find little else but epidemiology, and causal epidemiology at that. The reason why person *A* believes one thing and *B* believes another is simply and solely that *A* was born on one continent and *B* on another. Testability, evidential support, and the rest aren't even remotely considered. For scientific belief, epidemiology merely comes along afterward and describes the history of its acceptance. For religious belief, epidemiology is the root cause.

## Epilogue: Happily, Viruses Don't Win Every Time

Many children emerge unscathed from the worst that nuns and mullahs can throw at them. Anthony Kenny's own story has a happy ending. He eventually renounced his orders because he could no longer tolerate the obvious contradictions within Catholic belief, and he is now a highly respected scholar. But one cannot help remarking that it must be a powerful infection indeed that took a man of his wisdom and intelligence—now president of the British Academy, no less—three decades to fight off. Am I unduly alarmist to fear for the soul of my six-year-old innocent?

## Notes

1. R. Dawkins, *The Extended Phenotype* (Oxford: W. H. Freeman, 1982).
2. D. C. Dennett, in press.
3. G. C. Williams, "Pleiotropy, Natural Selection, and the Evolution of Senescence," *Evolution* 11 (1957): 398–411.
4. H. Thimbleby, "Can Viruses Ever Be Useful?" *Computers and Security* 10 (1991): 111–14.
5. R. Dawkins, *The Extended Phenotype.*

6. R. Dawkins, *The Selfish Gene* (Oxford: Oxford University Press, 1976).

7. See "On Viral Sentences and Self-Replicating Structures" in D. R. Hofstadter, *Metamagical Themas* (Harmondsworth: Penguin, 1985).

8. A. Kenny, *A Path from Rome* (Oxford: Oxford University Press, 1986).

9. Sir T. Browne, *Religio Medici I*, 9 (1635): 11.

10. A. Zahavi, "Mate Selection—A Selection for a Handicap," *Journal of Theoretical Biology* 53 (1975): 205–214.

11. R. Dawkins, *The Selfish Gene* (1976).

12. A. Grafen, "Sexual Selection Unhandicapped by the Fisher Process," *Journal of Theoretical Biology* 144 (1990): 473–516; "Biological Signals as Handicaps," *Journal of Theoretical Biology* 144 (1990): 517–46.

13. R. Dawkins, *The Selfish Gene*, 2nd ed. (Oxford: Oxford University Press, 1989).

14. D. C. Dennett, "Intentional Systems in Cognitive Ethology: The 'Panglossian Paradigm' Defended," *Behavioral and Brain Sciences* 6 (1983): 343–90; *Elbow Room: The Varieties of Free Will Worth Wanting* (Oxford: Oxford University Press, 1984).

15. M. Kilduff and R. Javers, *The Suicide Cult* (New York: Bantam, 1978).

16. A. Kenny, *A Path from Rome*, pp. 101–102.

17. Ibid., p. 101.

# Part III

# Social Issues

# 14

# The Church and the
# Martyr's Stake in Poland*

## Adam Michnik

The primary challenge confronting democracy in the post-Communist era is religious fundamentalism.

The controversy over the place of religion and of the church in a democratic country is the focal point of the debate over the form that democracy is to take in Poland. This question conceals an even broader issue: Will the utopia of totalitarian communism be replaced by a democratic order, a tolerant nation and a pluralist society; or will it be replaced by a new utopia based on religious or ethnic ideologies—the instruments of a new servitude?

The same kind of question arises in arguments over Islamic fundamentalism and over "ethnic cleansing" in the former Yugoslavia, in the conflict over religious parties in Israel, and in the discussions about about Protestant conservatism in the United States.

*Originally published in *New Perspectives Quarterly* 10, no. 3 (Summer 1993). Reprinted with permission.

## The Polish Church

In the days when there was neither a sovereign state nor any independent public opinion, when fear paralyzed any opposition and dictatorship believed that its power would never come to an end, the Catholic Church survived. For the polish people, the Church in the nineteenth century was a refuge from the Germanizing pressures of Protestant Prussia and the Russification policy of Orthodox Russia. As a result a real cultural entity arose, the "national-religious" identification of the Polish people. It endowed the Church with a sense of deep-rootedness and strength. At the same time, it gave rise to criticism of the Church from all sorts of non-conformist and rebellious spirits. The Church was the besieged fortress of the Polish sense of identity, and this identity was bound to Catholicism by a sacred tie.

Within this besieged fortress, surrounded by hostile forces, there was no room for sterile disputes regarded as stemming from petty cultural oddities. For the non-conformist and the rebellious, however, this atmosphere of a fortified camp, fixed in a belief from a conservative past, was nothing more than anachronistic provincialism, mental inertia and aggressive intolerance. It also meant that the Church was unable to carry on a dialogue with the contemporary world.

Thus a very special cultural group appeared in Poland. A sizable part of the elite, in the general sense of the word, withdrew from the Church and created a new center of ideas and culture. Anti-clericalism played a large role within this current.

Then came the Nazi occupation, followed by communism. The Catholic Church was the only institution which was able to maintain its independence during the totalitarian dictatorships. The Church's strategy—a dialectic of resistance and adaptation, of heroism and compromise—was wise and consistent. The Church survived the wave of repression, many trials and the imprisonment of Cardinal Stefan Wyszynski, the Primate of Poland. After his triumphant return from exile the Primate became, and remained for many long years, one of the main figures of Polish public life.

## The Polish Pope

The authority of the Church reached its apogee after the election of Cardinal Karol Wojtyla to the Holy See. Pope John Paul II's first visit to Poland, the strikes in August of 1980, the role of the Church under martial law all blended into a unique image of the spiritual and temporal power of the Church.

It took years for the intellectual elites to overcome their anti-clerical mindset. Their first meeting ground was the weekly *Tygodnik Forszechny*, edited by Jerzy Turowicz. Here could be found those individuals who were enemies of the Communist dictatorship but who also held themselves aloof from Catholic orthodoxy. Soon afterward, and at the same time that the elite's revolt against communism was growing, all those who were of any stature in the traditional Polish culture also began to appear in this weekly.

The monthly *Wiez*, edited by Tadeusz Mazowiecki, played a similar role. The encounters between the elites of the intellectual world and of the Church led to several reactions: This was an anti-totalitarian union in the name of national and democratic values; it was a recognition of the moral and spiritual role of the Church; finally, it was an expression of a renewed interest in religion and the metaphysical.

The Church at that time was seen as a friend, a partner in the dialogue. The Catholic periodicals reinforced this feeling with their wide-ranging content and their open spirit. The anti-clerical stereotype, it seemed, had been buried for all time. And then communism crumbled. In a common effort, with the Church taking a major role, the Round Table Accords (between the Communist government and the opposition which had rallied around Solidarity) were concluded, as were the elections of June 1990. The government of Tadeusz Mazowiecki was formed, and it was the dawn of a new era.

## The Crusades

A new tone could now be heard in the pronouncements of some of the bishops, and especially in the statements made by the politicians who were supported by these bishops. No more friendly tone, no more dialogue between partners. Now it was the voice of the Church, triumphant after a victory for which it claimed all the credit. The others were the ingrates, the enemies of the Church, of religion and of Christ—they disapproved of religious education in the schools; they opposed the criminalization of abortion and the bill which would include in the law on the mass media the obligation to "respect Christian values"; they criticized the active participation of the bishops in the election campaign, just as they protested the bishops' reticence during the anti-Gypsy pogrom in Mlava and the attacks on centers to care for the victims of AIDS.

The differences of opinion regarding the place of the Church in a democratic state are not surprising. And it is easy to understand the arguments about the constitutional and legal changes involved. The problem here, however, is that there was no argument; the dialogue had come to an end.

The bishops spoke the language of demands and rigid positions and met criticism with a tone of religious conflict, of a crusade. Those who criticized the demands of the Church were likened to the oppressors of the Communist regime, those who opposed the ban on abortion, to the defenders of the Holocaust. The entire debate over the form of democracy in Poland was cast in the language of the struggle between good and evil.

Of course, this description is oversimplified and applies to only a part of Polish Catholicism. A completely different tone permeated the statements of many bishops, whose ecumenical spirit outweighed the call to a crusade and whose open attitude was stronger than the spirit of fundamentalism. Yet the voice of fundamentalism was strong enough for it to become an essential element in the debate over the future of Poland.

For the Catholic fundamentalist, two forces had been battling for power in Poland: communism, with the support of Soviet bayonets; and Catholicism, deeply entrenched in the hearts of the Polish people. After the collapse of communism, therefore, power should of course be exercised by the Catholics who comprised the overwhelming majority of the population, enjoyed the support of the bishops and spoke the obvious truths since they were rooted in natural law and the teachings of the Church. For the Catholic fundamentalist, the words "I am a Catholic" have a special meaning in political discussions—for example, arguments over joining the institutions of the European Community—but only the fundamentalist is entitled to judge the authenticity of the statement.

Because indeed there exists, alongside the "real Polish Catholics," a "Catholic left" which, so says the fundamentalist, is a group that hides its leftist nature behind a Catholic vocabulary. "Leftism" for the fundamentalist, however, is a stain that can never be expunged: Anyone who has ever been in contact with the left and today does not accept the fundamentalist vision of a "Catholic government for the Polish people" is a "leftist," even though he may declare a thousand times over that this is not so. Equally "leftist," in this view, would be a member of the Political Bureau of the PZRP ("POUP," the former Communist Party) and a Solidarity militant who spent seven years in a Communist prison. The stain disappears, however, if the former member of the PZRP joins the political party of the Catholic fundamentalists.

## The Political Church

The fundamentalist considers it his duty to lead the reconquest of Poland by the "true Catholics," to wrench the country from the hands of the "leftists." The issue does not involve only the people who make up the political elite, however. It also concerns the form of the state. In the opinion of the fundamentalist, legislative acts should not be the results of compromise between various currents in a pluralist political scene. The law must be the expression of Catholic doctrine. It is the Church that must make the decisions regarding the criminalization of abortion and the death penalty, the limits on free speech, the vision of the world and culture, and finally, the outcome of the elections to the Parliament.

In this way, although it has never really been made clear, a Church is taking shape as a specific supra-power with decision-making authority in practically all domains of life, not only public but private as well.

Moreover, Catholicism is becoming a means of bringing together the masses of people whose political community is defined by a common religious identity. A set of ideas is thus being created, a mass political movement, established to confront the dissidents and the heretics. Whoever joins this movement feels himself to be superior because he knows the meaning of the revealed truths, while they remain hidden from those on the outside.

The natural consequence of these conditions is a vision of democracy as the rigid adaptation of governmental standards to Catholic ideas. The state and its institutions become the tools with which to express religious values. The renegades cease to be partners in the dialogue and become marginal members of society, to be tolerated and pitied. And any thought of a secular state is naught but the work of the Devil.

What was communism? The fundamentalist would reply that it was the end result of a process of secularization, the triumph of Reason divested of religion, beginning with the humanism of the Renaissance, through the rationalism of the Age of Enlightenment, and ending with the atheism of Bolshevism. And if Bolshevism has lost the game, then along with it the principle of the secularization of the state and the lay spirit of reason have also lost. Therefore, anyone who supports secularism and laicism must also desire the return of atheistic communism. He thus becomes the natural enemy of the political movement organized under the banners of Catholicism.

Obviously I have given here only a simple outline of the fundamentalist's creed. Its essence lies in the conviction that it must proceed from a "Church that knows no limits." However, the question of the place that should be occupied by the Church in a democratic state and a pluralist society is precisely a question of boundaries, of the limits of the Church.

The Church, distributing its invisible gifts in a visible world and having no limits in the transcendental domain, must be restricted in the way it functions as an institution. When a Catholic bishop claims the right to impose an all-encompassing concept of the world on a pluralist society, he is in fact declaring the fundamentalist creed. That has already happened many times in history and each time such pretensions have encountered firm resistance. Poland today will not be an exception to this rule.

## A Religious State

As always, the debate continues today in Poland with respect to the limits surrounding the various spheres of life which are granted their individual autonomy. The Church can become involved in these matters in two ways: as a partner in the dialogue with the real world, or as an institution which, because it speaks in a monologue, imposes its own political and legal order on the whole of public life. By choosing the second course, the Church is reduced to the status of any political pressure group and a Catholic political mass movement becomes the tool for using the Gospels to shape today's reality.

But this process has another side to it. The Church itself actually becomes a tool in the hands of the "national Catholic" leaders of the mass movement. And the leaders, operating behind the scene, seem to be arrogantly convinced that it is God Himself who has granted to them the knowledge of the inevitable destiny of Poland.

Today, Polish Catholicism is deeply divided: Opposing the fundamentalist vision is an attitude which stresses the autonomy within the sphere of political relationships. In this second view, the Church does not have the right to support one solution or another, be it institutional or constitutional. At the same time, there is reason to warn of the danger of those who, in the name of an ideology professing to be scientific or religious, feel that they are authorized to impose upon others their own perception of truth and well-being.

"Christian truth," according to the Catholic critics of fundamentalism, "does not belong in this category. Since it is not an ideology, the Christian faith does not believe that it can grasp, within a rigid scheme, such a differentiated socio-political reality; it believes that in the course of history human life has been expressed in various ways, which are far from being perfect." In other words, a wide spectrum of views exists with regard to the relations between Church and political activity, and this is also confirmed by some citations from a papal encyclical. Indeed, the conflict between these points of view has its own history, already very long and most likely to continue for some time to come.

The heart of the present argument can be seen in the various possible

interpretations of another very well-known passage from a papal encyclical. It says:

> Today we are in the habit of saying that the philosophy and attitude
> appropriate to the democratic forms of political life are agnosticism
> and skeptical relativism, while those who are convinced that they hold
> the truth and pursue it with determination do not deserve our confidence
> as far as democracy is concerned, because they do not agree with
> the position that, on the one hand, it is the majority which decides
> what is the truth and, on the other hand, truth changes according
> to modifications in the political equilibrium.
>
>     It should be noted in this regard that, when no absolute truth
> exists to guide political action and determine its course, it becomes
> easy to make tools of ideas and convictions to serve the objectives
> determined by the government. History teaches us that a government
> without values is easily converted into totalitarianism, whether this
> be overt or covert.

There is most certainly a religious plane where these considerations
may be debated, but I do not feel competent to discuss it. There does,
however, also exist a political plane.

## Secular Truths

On this point, it must be clearly stated that, in reality, no majority
can decide what is truth, and no changing political situation should
have the power to decide it. It would be an interference in a special
preserve, the conscience of each individual. It is, however, precisely the
majority that—through its representatives in Parliament or through the
referendum process—should decide the country's laws. In doing this,
however, it must respect all the liberties and civil rights of the minorities.
In a democratic state, the political arena is not the place where truth
is made a concrete reality, but is rather the arena for reconciling among
various interests just how this should be done.

    The political arena is thus a place for conflict and compromise.
Transforming it into a battlefield between truth and falsehood threatens

to invite the invasion of all sorts of fundamentalisms and the destruction of the democratic order. Who, then, should decide once and for all just what should be this "absolute truth that guides political action"? The bishops? The Catholic parties designated by them? Who would it be in other countries, dominated by a Protestant or Orthodox religion, by Islam or Buddhism? And how will this "absolute truth" be guaranteed? By censorship, the police, the penal code? We must share with others the fear that ideas can be converted into tools by a political authority.

And yet, what is this "absolute truth" that is truly protected from such a danger? It was certainly with the rhetoric of good Catholic truths that certain dictatorships have been upheld (we need mention only the Franco regime in Spain).

What is the meaning of a "democracy without values"? Is not the democratic system by its very nature the institutionalization of relativism and uncertainty, the art of forcing heterogeneous and conflicting entities to live together: peoples, nations, religions?

This principle of living together is regularly attacked by a kind of religiosity which attempts to turn its own religious standard into the standard of a state's law. Religious parties in Israel, Muslim or Catholic fundamentalism—there are many examples of political-religious movements that, in the name of the rights and principles of a pluralist democracy, began by demanding absolute tolerance for them to build their religious communities according to their own rules, only to attempt later, in the name of their own rights and principles, to impose their rules on everyone else.

## The Art of Limits

Democracy, on the other hand, is the art of self-limitations, the willingness of the majority not to impose its rule on others, in the name of diversity and of its own security. In this way, when the majority finds itself again in the minority, it will not see its own rights threatened.

Moreover, democracy does not specify other values than those which it guarantees by the very fact of their existence. All the rest must be the work of men and women, or of their communities which are organized to propagate certain values.

It is essentially thus that I view the truths of the democratic order. And yet, the passage from the papal encyclical on the movement from a "democracy without values" to "overt or covert totalitarianism" comes close to the actual experiences of the present day. Efficient and wealthy though they may be, the democratic nations are faced with the threat that their collective moral standards are disintegrating.

The ability of the community to defend itself in a democratic manner against the barbarian invasion of a "revolutionary utopia" or a "xenophobic authoritarianism" is weakening. Neither the democratic institutions of the state founded on law—including the educational systems— nor the principles of the market economy seem able to fend off the invaders.

It is not likely, however, that we can build our defenses by calling on the domination, legally established, of "absolute truth." Democracy is a risk. It is a daily plebiscite, offering a choice between the true and the false, between liberty and oppression, compromise or a debilitating civil war. Whoever seeks to remove this threat from the realm of the democratic order must inevitably destroy this order as well. Risk, it bears repeating, is an immutable feature of modern democracy.

## Taboos or Rule of Law?

At the beginning of his critical considerations of the pitfalls of the modern world, the Polish philosopher Leszek Kolakowski places much emphasis on the disappearance of the taboo.

But why should the taboo—that realm of attitudes protected by a tradition that resists rational argument—be respected? Why observe the precepts of kosher food for the Jewish people, the ban on alcohol for the Muslims, the Friday fast for the Catholics?

The various traditional bonds of mankind, explains Kolakowski, ensure the life of the community and without them our existence would be ruled solely by fear and greed. These bonds have no chance of surviving without a system of taboos, and it is perhaps better to believe in a few of them, foolish though they may seem, than to let them disappear.

To the extent that reason and rationalization threaten the existence of the taboo in our culture they weaken the culture's ability to survive.

In this situation, is it ultimately wise to resort to the uncertain hope that society's instinct for survival will prove to be sufficiently powerful that it will not let its taboos disappear? Is our social survival to rely on a hope that some new barbarians won't arrive in a reactionary way to enforce these taboos threatened by reason?

A democratic state is founded on law. Everything that is not prohibited in such a state is authorized. But "authorized" means only that it is not pursued by the police. For example, just because there may be no follow-up to an accusation does not mean that it should not be subject to moral opprobrium. There are moral and cultural standards which do not allow it.

Daily life is thus governed not only by the constitution and the penal code, but also by extra-legal standards. These standards ensure that society is not simply an assemblage of individuals, unknown and hostile to each other, but a community living in its "common home," using a common set of signals, customs, and values. This community's awareness of the rules governing life, even though fraught with conflict, enables each one of us to feel like a creature endowed with natural dignity and not an animal driven solely by the search for food and physical shelter.

In Poland, the Catholic religion is just such a factor in the idea of "a creature in his own home." The Christian system of values, as well as the attitude, the culture and the ethics which they engendered, are irreplaceable fixtures in the Polish home.

This, however, is not the entire home. Another of its fixtures is the concept of a "nation with no martyr's stake," of a tolerant nation which treats religious and cultural pluralism as a sign of health.

Still another fixture of this home is the spirit of revolt, the rush to defend Galilee, that same spirit so stigmatized by the Catholic inquisitor. It was indeed a Polish "dissident" who defended Galilee without waiting for the Apostolic Capital to render the verdict of rehabilitation. And yet, even while objecting to the judgment of the Catholic inquisitors, the "Polish dissident" neither wished nor sought to denigrate the Catholic Church, so great was his belief that the evangelic message was the spiritual property of the entire community.

Any attempt to institutionalize Catholic standards in a system of legal constraints regulating the life of the "common home" goes against the very idea of an open society. It is a form of political fundamentalism and leads to dictatorship.

Paradoxically, forcing the community into a mold of legal norms would lead to nihilism and the self-destruction of an open society. The "common home" would then be transformed into a ruin with neither roof nor door and the panes of its windows shattered.

These are the observations that we can make with absolute certainty. The rest is a stubborn effort to protect the values of the community in the world of risk which leads to liberty.

# 15

# The Rebirth of Democracy in Russia*

## Elena Bonner

## The April Referendum

Russia has been going through one more critical stage of development in the difficult transition to democracy, a stage that is at once cause for optimism and pessimism. Key members of the Congress of People's Deputies did their utmost to ruin the April 1993 national referendum that was in essence meant to determine the fate of the policies and presidency of Boris Yeltsin. Specifically, they attempted to rig the questions on the referendum ballot so as to ensure a vote of no confidence. But their efforts failed spectacularly, and once again the Russian people unequivocally demonstrated their loyalty to President Yeltsin and to the cause of democracy.

So much for the good news. The serious problem of deteriorating relations between the Congress of People's Deputies and the president remains. The current deputies were elected in March 1990, when the Communist Party was still in power. Some "experts" have claimed that they were elected by fair and democratic means, but this is not true.

*Originally published in *Imprimis* (Hillsdale College, Mich.) 22, no. 8 (August 1993). Reprinted with permission.

As a result, the overwhelming majority are old Party functionaries and members of the *nomenklatura*. Sixty-two percent—i.e., 639 out of a total of 1,033 deputies—consistently oppose democratic reforms. Just before the referendum, 618 actually voted to impeach Yeltsin. Only 38 percent—394 deputies—consistently support the president and the policies of reform. Each group spends most of its time battling to win over wavering deputies. In this environment, it is highly unlikely that the Congress of People's Deputies can achieve any substantive reform.

## An Anticonstitutional Crisis

Russia desperately needs—and needs soon if more violence is to be averted —a new written constitution. Without one, we will see more of what happened in the streets of Moscow on May 1, 1993, when deputies upset by the outcome of the referendum incited massive street violence in Moscow —the likes of which hasn't been seen since 1917 when the Bolsheviks used the same tactics in trying to come to power. In this case, tragically, over 500 hundred people were injured and one person was killed.

Democratic, pro-constitutional forces squandered their last political victory in August 1991 after the failed coup attempt when it would have been feasible to painlessly adopt a new constitution and to change the membership of the Congress. They must not squander their victory now. Two or three months ago, you could not have drummed up much interest in a new constitution, but now, after the successful referendum, it is on everyone's mind. On April 29, 1993, the Yeltsin government unveiled its proposed version of a new constitution. It seems to be the most democratic and the most adequate response yet to the needs of the nation. In my opinion, it still gives too much power to the president, but this can be addressed.

The first and primary chapter in the Yeltsin constitution guarantees the civil rights of all citizens. The second chapter outlines a federalist system in which autonomous republics, regions, provinces, and local governments retain a large degree of independence. (Anti-reform elements in Congress strongly oppose this provision. They would rather follow the old Soviet model of centralized power. But Yeltsin is adamant that the only way to save Russia is to allow decentralization.) In addition, the Yeltsin constitution calls for a whole new structure for the national government, featuring a two-chamber parliament with wide representation and four-year term limits. It also guarantees the inviolability of private property rights, including land ownership.

The constitution Russians are forced to live under right now is a relic of communism. It was written in 1936, and for decades it was known simply as the "Stalin Constitution." Then in 1978, when it was revised to further tighten the grip of the Communist Party, it was dubbed the "Brezhnev Constitution." During the last year and a half, 342 amendments to the "Brezhnev Constitution" have been passed by the Congress of People's Deputies, but this has only succeeded in making matters more confusing and contradictory and has forestalled any genuine improvement. Instead of serving as the supreme law of the land, the constitution is still the instrument of self-serving politicians. Some observers, therefore, have characterized this stage of Russia's development as a constitutional crisis, but in reality it has been an anticonstitutional crisis.

## The Nuclear Arms Issue

There is another crisis looming on the horizon for Russia. In an interview a few days before the April 25 referendum, Ruslan Khasbulatov, the chairman of the Supreme Soviet of the Congress of People's Deputies, was asked by the Western press about the ratification of the START agreements. He replied categorically that the Congress would not ratify any arms treaty until Andrei Kozyrev, minister of foreign affairs and one of Yeltsin's staunchest supporters, was fired or forced to resign. In other words, arms reduction has become a hostage that can be ransomed only for a certain political price. The Congress is filled with deputies like Khasbulatov who think and behave this way. They display a deadly combination of infantilism, belligerence, and irresponsibility that Western leaders should heed, especially when they call for all Soviet nuclear weapons to be transferred to Russia. Until Russia becomes a stable democratic state with leaders who pledge to abide by the law rather than their own whims, no weapons should be transferred. Just imagine for a moment that the people who were behind the May 1 violence in Moscow suddenly had total political power backed up by total control of the only nuclear arsenal in the Commonwealth of Independent States (CIS).

## Russia and the CIS

Some say that the real hope for peace and progress lies in once again uniting all the former Soviet republics under the banner of one government. But their ethnic roots, histories, and cultures are far too different. Nothing short of World War III would ever unite them again. But the new Russian constitution could be an enormous benefit for all CIS countries. Leonid Kravchuk, president of the Ukraine, acknowledged as much when he endorsed Yeltsin just before the April referendum. The fact that he chose to make his support public marks a watershed in the post-communist era, for up until now CIS solidarity has been a sham. This unprecedented overture has signaled that a new era of cooperation between CIS countries has begun.

## Western Aid

There is one more issue that I want to mention, the issue of Western aid. First, it is vital that this aid be distributed equitably to all CIS countries. Russia should not receive a disproportionate share; she is not, contrary to what you may see in the news, on the verge of starvation. But there are regions in the former USSR that are hunger-stricken. These are the ones that are caught up in armed conflict such as Tadgikistan; or the ones that have been devastated by natural disaster like Kyrgyzstan, which suffered an earthquake that destroyed the last harvest. There is also Armenia, which has been subjected to blockade since 1989; Ingushetia with thousands of homeless as a result of conflict with Osseria; and Abkhasia, which is in need of aid because of its ongoing war with Georgia.

Second, the Jackson Amendment of the 1970s should be revived.

No U.S. aid should be given to countries where human rights are routinely violated. Other nations should follow this example when formulating their own aid policies. Unless aid is linked directly to human rights, the West has no leverage to effect change—it is only subsidizing injustice and tyranny.

Third, Western aid should not be the most important or the only method of helping. Money, even when it amounts to billions of dollars, cannot overcome every problem. Sometimes it can even make problems worse. If the West really wishes to help, it should support efforts in CIS countries to establish democratic constitutions that will guarantee human rights, a stable currency, private property, foreign investments, free trade, and the rule of law. Western creditors should also consider postponing debt payments, especially since the debts in question were incurred by Yeltsin's communist predecessors.

## The Generation That Is the Future

I said at the outset that this stage in Russia's transition to democracy is cause for optimism and pessimism. Ultimately, I think optimism will triumph. Why am I so sure? It is not just because of the huge turnout for the April 1993 referendum, even though that turnout was phenomenal by any standards. It is mainly because I have seen *who* turned out. The biggest pro-Yeltsin, pro-reform group was comprised of Russian men and women between twenty and thirty-five years old. These young people are better educated and better trained than ever before and they have something that is totally new in Russian society: a global mentality. Moreover, they outnumber those who oppose reform—the retired, the veterans of war, of labor, and of the Communist Party. These young people *are* Russia's future. They will not give up on freedom and we should not give up on them.

# 16

# Real American Patriots Ask Questions*

## Carl Sagan and Ann Druyan

A few years ago, we were at a dinner in Peredelkino, a village outside Moscow where Communist Party officials, retired generals, and a few favored intellectuals have their summer homes. The air was electric with the prospect of new freedoms—especially the right to speak your mind, even if the government doesn't like what you're saying.

But, despite *glasnost,* there were widespread doubts. Would those in power really allow their critics to be heard? Would freedom of speech, of assembly, of the press, of religion, really be permitted? Would people inexperienced with freedom be able to bear its burdens?

Some of these Soviet citizens had fought—for decades and against long odds—for the freedoms that most Americans take for granted; indeed, they had been inspired by the American experiment, a real-world demonstration that nations could survive and prosper with these freedoms intact. They even were considering the possibility that prosperity was *due* to freedom—that, in an age of high technology and swift change, the two rise or fall together.

There were many toasts, as there always are at dinners in the USSR.

---

*© Carl Sagan and Ann Druyan. Originally published in *Parade,* September 8, 1991.

The most memorable was given by a world-famous Soviet novelist. He stood up, raised his glass, looked us in the eye and said, "To the Americans. They have a little freedom." He paused, then added: "And they know how to keep it."

Do we?

Science may be hard to understand. It may challenge cherished beliefs. In the hands of politicians or industrialists, it may lead to weapons of mass destruction and grave threats to the environment.

But one thing you have to say about science: It delivers the goods. If you want to know when the next eclipse of the sun will be, you might try magicians and mystics, but you'll do much better with scientists. They can tell you within a fraction of a second when an eclipse will happen decades or centuries in the future, how long it'll last, and where on Earth you should be standing to get a good look. If you want to know the sex of your unborn child, you can consult astrologers or plumb-bob danglers all you want, but they'll be right, on average, only one time in two. If you want real accuracy, try science.

What is the secret of its success? Partly, it's this: There is a built-in error-correcting machinery. There are no forbidden questions in science, no matters too sensitive or delicate to be probed, no sacred truths. There is an openness to new ideas combined with the most rigorous, skeptical scrutiny of all ideas, a sifting of the wheat from the chaff. Arguments from authority are worthless. It makes no difference how smart, august, or beloved you are. You must prove your case in the face of determined, expert criticism. Diversity and debate between contending views are valued.

Scientific findings and theories are routinely subjected to a gantlet of criticism—oral defenses of doctoral theses, debates at scientific meetings, university colloquia punctuated by withering questions, anonymous reviews of papers submitted to scientific journals, refutations and rebuttals. There is a reward structure built into science for finding errors: The more basic and fundamental the error exposed, and the more widely accepted it was, the greater is the reward.

This may sound messy and disorderly. In a way, it is. Science is far from perfect. It's just the most successful method known, by far, to understand the world. The discipline of science is hard; scientists, being human, don't always follow the methods of science themselves. Like other people, they don't especially enjoy having their favorite ideas challenged. But they recognize it as the cost of getting to the truth. And the truth—rather than the confirmation of their preconceptions—is what they're after.

Wherever possible, scientists experiment. They do not trust what is intuitively obvious. That the Earth is flat was once obvious. That heavier bodies fall faster was once obvious. That blood-sucking leeches cure disease was once obvious. That some people are naturally and by divine decree slaves was once obvious. That the Earth is at the center of the universe was once obvious. The truth may be puzzling or counter-intuitive; it may contradict deeply held prejudices. But, as the history of both science and politics has amply demonstrated, preferring comfortable error to the hard truth is, sooner or later, disastrous.

At another dinner in another city, many decades ago, the physicist Robert W. Wood was asked to respond to the toast, "To physics and metaphysics." By "metaphysics," people then meant something like philosophy, or truths you could recognize just by thinking about them. Wood answered along these lines:

The physicist has an idea. The more he thinks it through, the more sense it seems to make. He goes to the scientific literature, and the more he reads, the more promising the idea seems. Thus prepared, he devises an experiment to test the idea. The experiment is painstaking. Many possibilities are checked. The accuracy of measurement is refined. At the end of all this work, however, the idea is shown to be worthless. The physicist discards it, frees his mind from the clutter of error and moves on to something else. The difference between physics and meta-physics, Wood concluded, is that the metaphysicist has no laboratory.

Now, humans are not electrons or laboratory rats. But every act of Congress, every Supreme Court decision, every National Security Directive of the president is an experiment. Every change in economic policy, every increase in funding for Head Start, every toughening of criminal sentences is an experiment. Communism in Eastern Europe, the Soviet Union, and China was an experiment. Japan and West Germany investing a great deal in science and technology and next to nothing on defense was an experiment. Ideas can be tested.

Some of the opponents of the U.S. Constitution insisted that it would never work; that a republican form of government spanning a continent with "such dissimilar climates, economies, morals, politics and peoples," as George Clinton of New York said, was impossible; that such a government and such a Constitution, as Patrick Henry of Virginia declared, "contradicts all the experience of the world." The experiment was tried anyway.

Even a casual scrutiny of history reveals that we humans have a sad tendency to make the same mistakes again and again. Having power tends to corrupt us. We're afraid of strangers or anybody who's a little different from us. When we get scared, we start pushing people around. We have readily accessible buttons that release powerful emotions when pressed. We can be manipulated into utter senselessness by clever politicians. Give us the right kind of leader, and we'll gladly do just about anything he wants—even things we know to be wrong. The framers of the Constitution were students of history. In recognition of the human condition, they sought to invent a machine that would keep us free in spite of ourselves.

Thomas Jefferson, who had written the Declaration of Independence, had little to do with the Constitution; as it was being formulated, he was abroad, serving as America's ambassador to France. When he read its provisions, he was very pleased, with two reservations: One was that there was no limit on the number of terms the president could serve. This, Jefferson feared, was a way for a president to become a king, in fact if not in law. The other major deficiency was the absence of a bill of rights. The citizen—the average person—was insufficiently protected, Jefferson thought, from the inevitable abuses of those in power.

In 1791, with James Madison playing a leading role, the first ten amendments were added to the Constitution. From the beginning, they were called the Bill of Rights. The Constitution, the Bill of Rights, and the subsequent amendments were a new force in the world—a government with a written set of rules that distribute political power among the various segments of the society, with specific machinery not just vaguely allowing but actually protecting the expression of unpopular ideas. It was a machine of government designed to correct itself—like the evolutionary process, adapting to changing environments.

The Constitution is a daring and courageous document because it allows for continuous change—even of the form of government itself—if the people so wish. Because no one is wise enough to foresee which ideas may prove useful and answer urgent societal needs—even if they've been counterintuitive and disquieting in the past—it guarantees the fullest and freest expression of views.

There is, of course, a price. Most of us are for freedom of expression when there's a danger that our own views will be suppressed. We're not all that upset, though, if views we hate encounter a little censorship here and there. But within certain narrowly circumscribed limits—Justice Holmes' example was causing panic by falsely crying "fire" in a crowded theater—great liberties are permitted in America:

• Even if they mock Judaeo-Christian-Islamic values—even if they ridicule everything most of us hold dear—devil-worshipers (if there are any) are entitled to practice their religion, so long as they break no law.

• Those who oppose abortion are free to picket the abortion clinics and to display pictures of mangled fetuses, as long as their protest remains peaceful and nonobstructive. The fact that those who come to have abortions understandably do not wish to see these pictures is insufficient grounds for banning the demonstration.

• A purported scientific article asserting the "superiority" of one race over another may not be censored by the government, no matter how pernicious it is; the cure for a fallacious argument is a better argument, not the suppression of ideas.

• Individuals are free, if they wish, to praise the lives and politics of such undisputed mass murderers as Adolf Hitler and Josef Stalin. Even detestable opinions have a right to be heard. And we are free to contest them, or ignore them.

When permitted to listen to alternative opinions and engage in substantive debate, people have been known to change their minds. It can happen. For example, Hugo Black, in his youth, was a member of the Ku Klux Klan; but he later became a Supreme Court justice and was one of the leaders in the historic Supreme Court decisions, partly based on the Fourteenth Amendment, that affirmed the civil rights of all Americans. It was said that when he was a young man, he dressed up in white robes and scared black folks; and when he got older, he dressed up in black robes and scared white folks.

In matters of criminal justice, the Bill of Rights recognizes the temptation that may be felt by police, prosecutors, and the judiciary to intimidate witnesses and expedite punishment. The criminal-justice system is fallible: Innocent people might be punished for crimes they did not commit; governments are perfectly capable of framing those who, for reasons unconnected with the purported crime, they do not like. So the Bill of Rights protects defendants. A kind of cost-benefit analysis is made. The guilty may on occasion be set free so that the innocent will not be punished. This is not only a moral virtue; it also inhibits the misuse of the criminal-justice system to suppress unpopular opinions or despised minorities.

New ideas, and creativity in general, always represent a kind of freedom —a breaking out from previous hobbling constraints. Freedom is a prerequisite for science—which is one reason the Soviet Union could not remain a totalitarian state and be competitive in science and technology.

The Constitution was, in a way, a product of the scientific revolution. Once you questioned the prevailing religious insistence that the Earth did not turn, why should you accept the repeated assertions by many religions that God sent kings to rule over us? In the seventeenth century, it was easy to whip English and Colonial juries into a frenzy over this impiety or that heresy. They were willing to torture people to death for their beliefs. By the late eighteenth century, they weren't so sure.

The Bill of Rights decoupled religion from the state, in part because so many religions were steeped in an absolutist frame of mind—each convinced that it alone had a monopoly on the truth and therefore eager for the state to impose this truth on others. Often, the leaders and practitioners of absolutist religions couldn't see the middle ground or realize that the truth might draw upon and embrace apparently contradictory doctrines.

The framers of the Bill of Rights had before them the example of England, where the ecclesiastical crime of heresy and the secular crime of treason had become nearly indistinguishable. Many of the early

Colonists had come to America fleeing religious persecution—although some were perfectly happy nonetheless to persecute other people for *their* beliefs. The Founders recognized that a close relation between the government and any of the quarrelsome religions would be fatal to free inquiry.

Now, it's no good to have such rights if they're not used—a right of free speech when no one contradicts the government, freedom of the press when no one is willing to ask the tough questions, a right of assembly when there are no protests, universal suffrage when less than half of the electorate votes, and so on. Use 'em or lose 'em.

Due to the foresight of the framers of the Bill of Rights, it's hard to bottle up free speech. School library committees, the immigration service, the police, the FBI—or the ambitious politician looking to score cheap votes—may attempt it from time to time, but sooner or later the cork usually pops. The Constitution is, after all, the law of the land, and public officials are sworn to uphold it. However, through lower educational standards, declining intellectual competence, diminished zest for debate, and social sanctions against skepticism, our liberties can be eroded and our rights subverted.

The Founders understood this well: "The time for fixing every essential right on a legal basis is while our rulers are honest, and ourselves united," said Thomas Jefferson. "From the conclusion of this [Revolutionary] war we shall be going downhill. It will not then be necessary to resort every moment to the people for support. They will be forgotten, therefore, and their rights disregarded. They will forget themselves but in the sole faculty of making money, and will never think of uniting to effect a due respect for their rights. The shackles, therefore, which shall not be knocked off at the conclusion of this war will remain on us long, will be made heavier and heavier, 'til our rights shall revive or expire in a convulsion."

Education about the value of free speech and the other freedoms reserved by the Bill of Rights, about what happens when you don't have them, about how to exercise and protect them, is an essential part of being an American. If we can't think for ourselves, if we're unwilling to question authority, then we're just putty in the hands of those in power.

But if the citizens are educated and form their own opinions, then those in power work for *us*.

John Stuart Mill warned of a society letting "any considerable number of its members grow up as mere children, incapable of being acted on by rational consideration of distant motives." Jefferson said, "If a nation expects to be both ignorant and free in a state of civilization, it expects what never was and never will be." He continued the thought in a letter to Madison: "A society that will trade a little liberty for a little order will lose both, and deserve neither."

We suggest that a good way to celebrated this 200th anniversary of the Bill of Rights is to rededicate ourselves to the habits of free inquiry, skeptical scrutiny, exposure of government actions to public view, and support for the right to express all opinions—including, especially, those we find personally distasteful. In a democracy, opinions that upset everyone are sometimes exactly what we need. We should be teaching our children the scientific method and the Bill of Rights.

We Americans were once legendary for our inventiveness. We were considered disarmingly original, audacious, able to do the impossible. This is, by and large, not a description of the United States today. We must make sure our children understand why the vigorous exercise of our civil liberties is essential—not just to retain what freedoms we have, freedoms celebrated and envied in countries we were once taught to think of as our adversaries, but also to preserve and invigorate our nation's soul.

# 17

# Science and the Making
of Society in Latin America*

## José Leite Lopes

What is our physical image of the world today? How was it achieved throughout history?

What was the importance of scientific research, of cultural creation, for the development of contemporary advanced societies?

What was the role of the lack of scientific education and practice on underdevelopment? And, in particular, for Latin America?

What are the present-day problems, what are the prospects for the application of science and technology to promote adequate development not only in less developed countries but also in advanced industrial nations?

Is the aim of science and technology to liberate man or to contribute to a world ruled by repression of the many poor by the few rich?

These are some of the questions which we all must study and try to answer if we are to formulate meaningful proposals for a harmonious development of human societies in accord with their cultural heritage and national identity in the changing world of today.

---

*Originally published in *Interciência* 5, no. 3 (1980): 159–65. Reprinted with permission.

# The Physical Image of the World

As is well known, speculations about the structure of the universe were always included in the cosmogonic models and philosophical systems developed by ancient civilizations.

In Asia, in Africa, in Latin America, superb achievements were obtained by ancient societies—in their mythical approach to the study of nature, in their cultural monuments, in their artistic and technological ingenuity, in their astronomical observations, in the philosophy of space, time, matter, and life that they created and which reflected their modes of interactions with the world.

It was, as everyone knows, the atomistic philosophers of ancient Greece who exercised perhaps the greatest influence on the modern conception of the universe.

Before the Greeks, the Babylonians and the Egyptians had already made observations, during many centuries, on the motions of the sun and of the moon with respect to the fixed stars, and they knew how to predict lunar and solar eclipses. Even though the Greeks assimilated celestial bodies to gods, Anaxagoras stated that the sun was like a red hot stone and that the moon was made like the Earth. The Pythagoreans, at the end of the fifth century B.C., stated that the Earth is spherical; Aristarchus of Samos, in the third century B.C., discovered the complete Copernican system, and Eratosthenes, in the year 200 B.C., calculated, according to Ptolemy, the maximum distance between the moon and the Earth and the minimum distance between sun and Earth.

Abu 'Ali al-Husayn ibn 'Abd-Allah ibn Sina, known as Avicenna, philosopher, codifier of Aristotle, and one of those who preserved Greek culture and contributed to its transmission, stated: Time is the measure of motion.[1] In the Rasa'il, a fifty-one-treatise encyclopedia known as the Koran after the Koran, one finds a list of distances to the planets (as a function of Earth radii) and of sizes of planets; it is stated there that space is "a form abstracted from matter existing only in the consciousness."[2] But how many documents were lost or destroyed, as happened for instance in the subjugation of the magnificent pre-Colombian civilizations by the invading Spaniards in Mexico and in Central and South America?

After these systems were forgotten during the decay of later antiquity, the Christian medieval model of the image of the world arose. As expressed in Dante's *Paradiso,* the Earth is the center of the universe, Satan is at the center of the Earth, the heavens consist of ten concentric spheres. Everything below the moon is subject to corruption and decay; everything above the moon is indestructible. "God, the Aristotelian Unmoved Mover, causes the rotation of the Primum Mobile, which, in turn, communicates its motion to the sphere of the fixed stars and so on downward to the sphere of the moon."[3]

The great scientific revolution in astronomy and in physics came long after the Greeks, in the sixteenth and seventeenth centuries, with the work of Galileo and Newton, who constructed the first scientific image of the universe.[4] By discovering the laws of motion of the bodies of our daily experience and by generalizing these laws to all bodies in the universe, and by inventing the infinitesimal calculus needed for this work, Newton achieved the first great synthesis and this—by ultimately correlating ideas and facts apparently strange to one another —has become the aim of modern science: the fall of an apple from a tree, the motion of the moon around the Earth, the motion of the celestial bodies under the action of universal gravitation. Einstein wrote in his autobiographical notes:

> In the beginning (if there is such a thing), God created Newton's laws of motion together with the necessary masses and forces. This is all; everything beyond this follows from the development of appropriate mathematical methods by means of deduction. What the nineteenth century achieved on the strength of this basis, especially through the application of the partial differential equations, was bound to arouse the admiration of every receptive person.[5]

After the Newtonian mechanics of action at a distance, the notion of field was introduced in physics, mainly through the work of Faraday and Maxwell on electromagnetism, which culminated with another great synthesis, unifying the domains of optics, electricity, and magnetism. What made Maxwell's theory "appear revolutionary," wrote again Einstein,[6] "was the transition from forces at a distance to fields as fundamental variables. In this connection I cannot suppress the remark that the pair Faraday-Maxwell has a most remarkable inner similarity with the pair Galileo-Newton—the former of each pair grasping the relations intuitively, and the second one formulating those relations exactly and applying them quantitatively."

At the end of the nineteenth century, the discovery of the electron and of the proton took place and a number of remarkable questions arose which led, on the one hand, to the discovery of the quantum of action by Planck in 1900 and, on the other hand, to the development of the theory of relativity by Einstein in 1905.

> When one looks back over the development of physics, one sees that it can be pictured as a rather steady development with many small steps and superposed on that a number of big jumps. Of course it is these big jumps which are the most interesting feature of this development. The background of steady development is largely logical, people are working out the ideas which follow from the previous set-up according to standard methods. But then, when we have a big jump, it means that something entirely new has to be introduced. These big jumps usually consist in overcoming a prejudice.[7]

The inventive physicist finds that he has to question this prejudice and replaces it by an entirely new image of nature.

In his work on the special theory of relativity, Einstein made one of these big jumps, achieving a new great synthesis of apparently disconnected ideas: the prejudice of absolute simultaneity was questioned, analyzed, and replaced by a new conception of physical space, a new entity in which ordinary three-dimensional space and time are amalgamated to constitute a four-dimensional manifold, a consequence of which is that space may generate time, energy may generate momentum, energy is equivalent to mass, electric and magnetic fields are aspects of the same subjacent variable, *the* electromagnetic field.

Moreover, a new concept, that of a *superlaw,* was introduced by Einstein in physics with his relativity principle. By postulating that the laws of physics must be independent of the state of (rectilinear and uniform) motion of the observer, of his position in space, and of the time at which his observations are made, Einstein formulated a general requirement to be satisfied by the equations of physics. If an ordinary physical law expresses a relationship between variables associated with phenomena and events, the principle of relativity states how such a relationship must be expressed, how it may not be; mathematically, the principle of relativity states, as is well known, that the physical laws must be invariant under a certain group of transformations, the Poincaré group. This was perhaps a striking and very precise realization of the claim, or desire, that scientific knowledge must be wholly impersonal, independent of the physicist who makes the experimental observations. And so was also the proclamation of absolute statements —the invariant laws—as well as the relativization of the notion of measure, of the values of length, volume, time interval, energy of a physical system, for example, as numbers which depend on the frame of reference in which the measure is carried out.

It was still Einstein who after ten years of research discovered the relativistic theory of gravitation, one of the most beautiful, if not the most beautiful, constructions in the theoretical physics of all time. By achieving a new synthesis, which generalized Newton's gravitation theory, Einstein identified the gravitational field with the tensor of the space metric, the physical space as described by laws of Riemannian geometry. The machinery of this geometry led Einstein to invent his equation of the gravitational field—an equation which is based on the notion that matter affects the curvature of space-time and that space-time acts back onto matter and determines the nature of its motion: a revolutionary concept which destroys the old notion of space as a passive stage where events take place, without affecting them, as proclaimed by Leibnitz.[8]

It was mainly his invention of the relativistic theory of gravitation which led Einstein to formulate his conception of the genesis of scientific knowledge in physics: the concept and the laws which relate these concepts to one another can be discovered by means of purely mathematical constructions, and give the key to the understanding of natural phenomena. Experience may suggest appropriate mathematical ideas but these can surely not be deduced from it. Experience, of course, remains as the only valid criterion for judging the physical utility of a mathematical theory. But "the concepts and principles are free inventions of the human intellect, which cannot be justified either by nature of that intellect or in any other fashion *a priori.*"[9]

This epistemological conception of scientific work, of so-to-say, an anti-Baconian character, is indeed to be found from Newton, Lagrange, Hamilton, to Einstein, De Broglie, Heisenberg, Dirac.

The discovery and the development of the theory which describes atomic phenomena—quantum mechanics—as well as research on the ultimate constituents of matter, the so-called elementary particles, has dominated the physics of the last fifty years.

It was only in the beginning of the eighteenth century that the atomic hypothesis, put forward by the Greek philosophers, lost its theological and metaphysical character. Democritus, in the fourth century B.C., stated that "the only existing things are atoms and the vacuum; all else is mere opinion"; and thereby put forward the important notion that the complex variety of bodies and phenomena result from the motions and interactions between invisible and indivisible particles, the atoms, which obey "simple" laws. Newton himself wrote:

> It seems probable to me that God in the Beginning formed matter in solid, massy, hard, impenetrable, movable Particles of such Sizes and Figures and with such other Properties and in Such Proportion to Space as most conduced to the End for which he formed them; and that these primitive Particles being Solids, are incomparably harder than any porous Bodies compounded of them; even so very hard, as never to wear out or break in pieces; no ordinary Power being able to divide what God himself made one in the first Creation.[10]

We all know that the atoms—or at least the objects we came to call atoms—were found to be rather complex systems. The development of modern science, from the seventeenth century to our days, the extraordinary achievements of experimental techniques and ingenuity, the birth and development of scientific thought led to the replacement of the metaphysical approach to natural philosophy by the rational approach based on experimental evidence, on mathematical models constructed on the results of observations and on inventive intuition.

The notion of indivisible atoms gave place to the concept of elementary particles and it was hoped that these particles would be—in small number—the fundamental constituents of matter. In the last decades, however, a large number of such sub-atomic particles were discovered, a number which is now in competition with the hundred atoms or so which integrate Mendeleyev's periodic table. We now know six species of *leptons*—particles which include—and have properties in common with—the electron and its neutrino. There are the *baryons,* particles related to protons and neutrons; the *mesons,* which are exchanged between baryons; there is the *photon,* the particle of light, responsible for the propagation of electromagnetic forces. We assume the existence of particles which have not yet been observed such as the *graviton,* which propagates gravity, the *weak mesons,* which propagate weak interactions. The mathematical beauty of the present attempts to unify the weak, strong, and electromagnetic forces, such as the Salam-Weinberg model, led most of the present day physicists to believe in the existence of the latter particles. This unification, in which the efforts of the physicists specialized in the domain of high-energy physics are concentrated, will constitute a new great synthesis, comparable to those mentioned earlier in this paper.

And this is the present hope: to reduce the different forms of observed forces, the gravitational interactions, the weak interactions, the electromagnetic forces, and the strong forces (responsible for the existence of nuclei and therefore of matter) to different manifestations of certain underlying basic entities called gauge fields.

This unification is an old dream which started with the attempts of Einstein to include the electromagnetic forces in the unification of gravitation and space-time geometry. And it is the method introduced with so much force and elegance by Einstein in theoretical physics, the search for symmetry groups which leave invariant basic physical laws, which is at the root of our present-day work.

Matter, on the other hand, the variety of elementary particles mentioned above, seems to be constituted—with the probable exception of the leptons—by certain objects called *quarks.* The quarks would be the latest elements in the fragmentation scalation of matter, a kind of ultimate particles which integrate the heavy elementary particles but which would be, for some reason not yet fully understood, not observed as a free particle. Quarks would most probably be confined inside the elementary particles and this notion would perhaps give the key for us to stop in the process of reduction of matter to smaller and smaller constituents.

To conclude this picture of our physical image of the world, let me say just a few words on the astrophysicist's conception of the universe.

It was after the discovery of the relativistic theory of gravitation that cosmology started to develop as a science. From 1917 on, observational data and theoretical studies laid the foundations of this discipline which has continued to develop ever since.

As stated by a distinguished cosmologist:

> The existence of the universe is clearly its most important characteristic, but I am referring here to the stronger idea that it is meaningful to talk of the universe as a whole, as a single well-defined concept. This idea is one of the most important, perhaps the most important, scientific discovery of the twentieth century.[11]

That the universe is unique and that we can apply to its study the physical laws which are established on this earth, are generally admitted postulates. Observations of stars and galaxies and objects revealed by the emission of invisible radiowaves have led astrophysicists to the conception that the whole universe is in a state of expansion, of continuous change with time. The notion of expansion of the universe was the result of observations of the red shifts of the galaxies, radio source counts, abundance of the elements. The extrapolation into the past of this outward movement of galaxies, the discovery of the background microwave radiation, have led to the conclusion that our world came into existence in a sudden way, out of an explosion, a "big bang." This is the so-called "standard hot big-bang model" according to which, at the beginning, about 15 billion years ago, elementary particles were highly concentrated and under thermodynamic equilibrium at extremely high temperature, with the decay of particles and recombination of pairs in the first few seconds. With the motion of these particles temperature dropped and elements were formed, with the production of helium out of protons and neutrons. Later on heavier elements were produced out of nuclear reactions and the stars became intensely hot furnaces in which ordinary matter was forged out of protons.

And thus we are still left with the question of what happened before the initial stages of nearly infinite density and temperature and pressure. Other models of the universe are also being investigated by cosmologists who ask fascinating questions such as the possible permanent expansion of the universe or its return to contraction, the gravitational collapse, the existence of singularities in nature.

Such is the evolution of our ideas about the universe, from the old civilizations to present days. It is fascinating to learn that the matter of our localized world, "the carbon and nitrogen of our bodies, the oxygen we breathe, the iron in our blood were all generated inside stellar furnaces at remote epochs in the past"[12]—and that in any case, as dreamed of by Anaxagoras in the times of Pericles, the sun is like a red-hot stone and the moon is made of earth.

## Science and Underdevelopment in Latin America

The above is only a sketch of some of the basic lines of our physical image of the world, the one with which I am familiar. Others might tell you about the foundations of the biologist's picture of the universe, of the points of view of the chemist, the geophysicist, the mathematician, the social scientist.

We see that modern science had its birth in the seventeenth century— and became associated with the emergence of capitalism in Western European countries. Little by little, empirical inventions of machines and mechanisms, the study of nature in the laboratory, the search for new products, and the understanding of the laws of nature furnished the instruments for the technological and scientific transformation of the world. At the same time, in other civilizations and societies, many of them subjugated by conquest and war, similar processes did not take place.

Immersed in a different historical context, subjected to specific religious, cultural, political, and economic forces, these societies did not develop the search for scientific knowledge—or were not allowed to continue such a development—and thus lacked basic tools for the transformation of the world and indeed for ensuring their very survival.

Once inequality among nations was established due to their different forms of interactions with the physical world, economic and political forces were bound to act in order to increase this inequality.[13] And the development of the emerging industrialized societies apparently arose associated with the economic and political domination of other societies, the present underdeveloped nations or, if you wish to use another, more emphatic nomenclature, the less developed countries.

It is certainly striking that Spain and Portugal did not greatly contribute to the great theoretical advances of science as did other countries of Western Europe. My remarks in the remaining of this paper suggest that the Iberian Catholic *Weltanschauung* was, among other factors, responsible for Iberian weakness in theoretical thinking, for the material and intellectual underdevelopment and dependence of Spain and Portugal and their former colonies. A corollary of this hypothesis is the suggestion that science of the highest quality must be supported in our part of the world as an intrinsic component of the process of social and economic development of this region.

In Latin America, as is known, Spaniard and Portuguese conquered the native people, and made efforts to destroy their religious systems and their cultural achievements. The civilizations in the Andes—the Incas—and in the Mexico plateau and in Yucatan—Aztecs and Mayas—had reached important levels of material and cultural development before the invasion in the first half of the sixteenth century.

Mathematics, which included the utilization of number zero, astronomical knowledge, which included the prediction of eclipses, were achievements of those civilizations which also developed techniques in agriculture, architecture, and engineering as well as an artistic culture which were transmitted from generation to generation.

The replacement of local cultures by those of Western Europe as brought about by Spain and Portugal did not lead, however, to scientific development in our part of the world, such as was taking place in parts of Europe.

It is true that Spain and Portugal made superb achievements, which culminated with the discovery of continents, such as the development of the art and sciences of navigation. Several factors, however, such as the great influence of religion and the power exercised by religious authorities in those two countries, prevented them from participating in the European creation of science in the seventeenth century.

It is not my purpose to describe the effect of this on the evolution of science in Latin America. Names and data can be found in books and specialized articles.[14] It will be seen that, in spite of difficult conditions of work, many talented scientists did important work in many countries of our continent, mainly after the second half of the nineteenth century. What is of the greatest interest to us is to see that the state of political and economic dependence of our countries did not allow the flourishing of science. The colonies of Central and South America were regarded as places rich in primary materials to be exported to the expanding capitalist countries of Europe. And these in turn exported to the Latin American colonies their industrial products. "[Latin America], it is not unjust to say, saved the British cotton industry in the first half of the nineteenth century, when it became the biggest market for the English exportations."[15]

The proclamation of political independence did not change the nature of the economic system in those countries—it was rather an opening toward their domination by Great Britain. At the same time, an ideology was taking shape which purported that the process of economic development was a kind of game, of free competition, where the most intelligent and dynamic people are successful. In fact, political and economic domination prevented other peoples from competing in these games.[16]

Within our countries' national ruling classes, affiliated with those in the dominating foreign powers, an ideology developed according to which our countries' role is to export raw material necessary to the expansion of capitalist industrialized countries.

"It was accepted," states Velho, "that we [in Brazil] would never be able to produce industrial goods so well as England and other countries and that if we attempted to do so and became protectionists we would certainly suffer retaliations against our agricultural exportations."[17]

Subsequently, the transformation of the economies of Latin American countries by industrialization, which started at the beginning of the twentieth century and developed in the 1930s according to the model of substitution of imports, had as a direct consequence the importation and the imitation of products and of the means of production invented abroad, as well as the purchase of technology developed in advanced countries.

It is thus important to emphasize that economic dependence, although under another form, was essentially kept untouched when Latin American countries ceased being formal colonies of Spain and Portugal. The search for manufactured products equal or similar to those which were imported led immediately to technological dependence from abroad—the scientific and technical knowledge necessary to industrialization in Latin America was incorporated in the machines and plants imported from abroad.

In parallel to this situation of the economy, the medieval Christian image of the world was imposed through education. Universities were founded late in Latin America—and exceptions do not affect the general feature of lack of scientific education and its subsequent effect on the life of our nations. Clearly, the absence of industries meant that there was no need for technological and scientific research institutes. And universities founded early in Latin America, in the seventeenth century, were rather centers dedicated to the study of the medieval-inspired culture developed in Spain and Portugal.

In the last few decades, a great effort has been made toward the development of universities and scientific institutes in many countries of Latin America. Following the industrialization process, many universities and scientific laboratories were founded or further supported and developed.

However, in spite of this expansion of the university system, of science and culture, the fact is that the industries which are owned by Latin American industrialists continue to depend basically on imported machinery and technology.

Our industrialists have never been concerned with the technological research necessary to the improvement of the quality of their manufactured goods. Associated with foreign enterprises, from which they buy equipment and technical assistance, national industries in Latin America seldom demand technical service from the national technological institutes.[18] In this way, Latin American universities have not generally been associated with studies for economic projects, and scientists and technologists have not been requested to help make fundamental decisions in the formulation of economic-development programs.

In advanced industrialized countries, on the other hand, machines and plants which are invented depend on intensive technological research and this, in turn, is based on fundamental science carried out in institutes and universities.

## Science and Dependent Development

If we Latin American scientists and research-engineers hoped one day to be able to contribute to the development of our countries, this hope was shattered by the government decisions taken in the last twenty years to base development on the implantation of affiliates of multinational enterprises. These industrial companies, which have their own research laboratories in the center of the capitalist system, produce goods in the countries where they establish themselves mainly for export or for consumption by a small fraction of our population. In recommending the adoption of these policies, technocrats utilize the myth of "technology transfer." The installation of plants of multinational enterprises clearly does not imply transfer of technical and scientific knowledge. Imported machines are invented, designed, and built abroad, and plans for making local goods cannot be modified by national engineers. Even if we do not consider the basic question of whether these industrial products are really those needed by our population, it is clear that the important objective is to create a capacity for technological innovation and not to teach workers which buttons to press. The capacity for technological invention is not transferred by multinational enterprises. Research is, therefore, carried out abroad and technology penetrates locked in black boxes.

The integration of most of Latin America into the economic-cultural market of the industrial capitalist nations has thus inevitably led to an aggravation of dependence: science and culture have become luxury imported products—sometimes locally produced by and for a few.

# Endogenization of Science in Which Society?

It is thus clear that if we are to discuss plans for the development of science and culture in our countries, if we are to suggest strategies for what is now called endogenization of the activities in these domains, we have beforehand to characterize clearly the political, economic, and social forces which have been obstacles to independent development, to the enrichment of our cultural heritage, to the affirmation of our national identities. For how are we to suggest a strategy for developing endogenous culture and science if the underlying political and economic systems prevent this endogenization, and assume that what has been invented abroad is necessarily what we must imitate and buy?

# The Aims of Science

As we follow the marvelous history of the elaboration of our scientific image of the universe, we are tempted to say that science is a unique and universal system of knowledge, politically neutral, and standing above ideologies.

Scientific laws are of course valid no matter which laboratory or which country they come from. But science is not only a catalogue of data, names, and statements. Scientific research is a dynamic process which includes interaction of the scientific community with their surroundings, with political and social forces. Motivations for research, its planning and funding are not politically neutral. Science, through its interpretation of the world, gives us as well instruments for changing it.

How many industries arose from pure fundamental research—from mechanics and thermodynamics, from branches of chemistry, from Maxwell's equations, from the theory of electrons and quantum mechanics? Is not the whole field of nuclear energy a result of, among other ingredients, the formula of equivalence between mass and energy?

Science, we have been taught—and we like to repeat it—works for mankind, for the benefit of man, for the liberation of man from work. Science and technology are indeed so powerful as to be able to send man into cosmic space. Could they, however, not be used to improve the living conditions of the poor and exploited masses in Africa, in Asia, in Latin America?

Science is indeed a part of the social and political system of industrially advanced nations. The results of scientific and technological research are then primarily and chiefly applied for the promotion of their model of society, for their own benefit.

There are those who do not attach importance to these questions, who elude them.

There are those who, confronted with political changes in developing countries which tend to liberate them from subjugation and dependence, pretend to tell these countries which kind of science they must develop. It is suggested by them, for example, that developing countries must utilize only so-called intermediate technologies, leaving the fields for advanced science and technology, the so-called "bid science," hard science, to the industrialized nations.

This suggestion is clearly unacceptable. Of course, a given country, with its specific resources, cannot always develop any chosen technology. Even the nations of Western Europe had to get together and pool their physicists, technicians, and financial means in order to establish a high-energy physics laboratory—CERN—so as to produce the advanced and expensive equipment needed for investigation of the ultimate structure of matter.

Is not this idea of getting together, of pooling human and material resources among nations of a given region of the world, a good idea, worthy of imitation? The capacity of developing countries will thus be enhanced and multiplied by a significant factor, fields of research in science and technology will open to groups of nations, each of which would be unable to develop them alone. Clearly, locally developed techniques, many of which tend to be replaced by imported and inadequate technologies, must be preserved and further studied.

The principle that developing nations must not have access to certain fields of knowledge is unacceptable—it would freeze the present division of the world into rich and poor nations, and perpetuate the international division of labor.

Of course, appropriate technologies, in the sense that they should be financially, economically, ecologically adequate, and serve the ideals of improving the living conditions of the whole community, not the interest of a privileged minority, are to be recommended not only to developing nations but also to the rich industrialized countries.

Energy is wasted, fossil fuels indiscriminately burnt in rich countries. Sophisticated equipment is indiscriminately exported by them to poorer countries just for their investments. Industries and services are automated where labor is available in enormous surplus, computers are installed everywhere with the subsequent aggravation of unemployment.

The aim of science and technology, under the capitalist system, would seem to be to liberate men from work—and thereby condemn them to unemployment without access to the basic requirements of life.

I believe, therefore, that it is meaningless to push forward strategies for scientific and technological development in our nations without at the same time analyzing and implementing a change of these countries' economic pattern.

We certainly cannot succeed in achieving endogenization of culture, science, and technology if we, developing nations' scientists, are not aware of the basic political and economic forces which have prevented—throughout our history—the development of our potential capacity for creation.

## Science for Liberation

Before concluding, let me still make a few additional remarks. The first remark concerns the relationship between the individual work of creation by scientists and the social and political significance of science and technology. It is quite clear that motivation for research work on the part of many scientists does not have any political or economic connotation. It is apparently the drive to understand and solve problems in their specialized field, associated with the description, correlation, and explanation of events and objects, and leading to the discovery of laws of nature and of new images of the world, that incites many scientists to do their research. In performing this work they take advantage of special intuition and develop a sense of beauty and elegance which only professional scientists know and which is probably not essentially different from feelings of creative artists, philosophers, writers.

The ensemble of scientific laws, theoretical views, and experimental techniques constitutes a body of knowledge of the physical world with strong interaction with technology—a transformation of science into the art of creating new products, of inventing machines and building mechanisms, which are the tools for transforming the world according to the needs of men and women, their political and social projects, and economic requirements.

Not only the great innovators contribute to the construction of science, but also lesser scientists. These scientists are also important in their search for details, consequences, and applications of fundamental new theories and experimental discoveries.

There are also those—probably the great majority—who do their research work according to specific programs and plans, with the idea of making discoveries of interest for a given practical domain, or specific fields such as solid state physics, electronics, nuclear energy, space physics, and so on.

The results arising from all this variety of research make up the various domains of science. It would not be correct to say that this ensemble is free from social, economic, and even political significance. Nuclear energy physics is perhaps the best known case where Einstein's fundamental work had a purely theoretical motivation, but where social, economic, and political meanings were far reaching—and at times ominous.

Contemporary science gives rise to all kinds of technologies which are responsible for social, economic, and political change in our world: from food production technology to the technology needed for the most dreadful and destructive weapon systems. Scientists are thus naturally led to think about social, economic, and political consequences of research, even if their own personal work involves only abstract ideas.

Scientists belonging to countries of the Third World, in particular, are naturally led to meditate on the role which science and technology may play in the making of their societies. They will find that discoveries made in the research laboratories and universities of advanced countries will be useful to their great industrial companies. These companies sell the result of their research work and then invest a part of their profits in the further development of these laboratories.

In a developing country whose economy is dominated by multinational enterprises, which employ almost exclusively their own scientific and technological knowledge, research work carried out in national research institutes and universities does not generally have application to the benefit of that country.

Developing nations' scientists can thus hardly escape the conclusion that the development of science and technology presupposes a political system whose aim will be the welfare of the whole population.

The following questions are thus appropriate before a science and technology policy is decided upon: which science and which culture, for which project of society in which world?

Is the aim of science and technology to liberate man or to establish a world ruled by repression of the many poor by the few rich?

To my mind, there can be no other answer: science must liberate men and women, and by that I do not mean men and women of the advanced societies. We must work for liberation of all men and women everywhere so that science will fulfill its vocation of universality and will become the patrimony of all mankind.

# Notes

1. C. W. Misner, K. S. Thorne, J. A. Wheeler, *Gravitation* (San Francisco: W. H. Freeman and Co., 1973), p. 753.

2. Ibid.

3. B. Russell, *Human Knowledge, its Scope and Limits* (London: George Allen and Unwin Ltd., 1948), p. 24.

4. The difference between the reputation of Aristarchus and that of Copernicus and Galileo lies, according to Russell, in the fact that in ancient Greece, astronomy was an amusement of the idle rich and not an activity integrated in the life of the community. In the sixteenth century, however, science had made important inventions, the discovery of the Americas had shown the limitations of the ancient knowledge of geography, Catholic orthodoxy had begun to be an obstacle to material progress and the fury of theologians made scientists appear as heroic champions of a new wisdom (B. Russell, *Human Knowledge*, p. 25).

5. P. A. Schilpp, ed., *Albert Einstein, Philosopher-Scientist* (Evanston,

Ill.: The Library of Living Philosophers, Inc., 1949), p. 19.

6. Ibid., p. 33.

7. P. A. M. Dirac, "Development of the Physicist's Conception of Nature," in *The Physicist's Conception of Nature,* ed. J. Mehra (Dordrecht, Holland: D. Reidel Publishing Co., 1973), pp. 1–14.

8. J. Leite Lopes, "The Evolution of the Notions of Space and Time," *Scientia* (Milan) 107 (1972): 411–33.

9. A. Einstein, *Ideas and Opinions* (London: Souvenir Press, 1973), p. 272.

10. I. Newton, *Opticks; Treatise of the Reflections, Refractions, Inflections and Colors of Light* (New York: Dover Publications, 1952—based on the 4th ed., London, 1730), p. 400.

11. D. W. Sciama, "The Universe as a Whole," in *The Physicists's Conception of Nature,* ed by J. Mehra (Dordrecht, Holland: D. Reidel Publishing Co., 1973), pp. 17–33.

12. F. Hoyle, *The Ten Faces of the Universe* (San Francisco: W. H. Freeman and Co., 1973), p. 78.

13. J. Leite Lopes, "Science for Developing a View from Latin America," *Bull. Atomic Scientists* 22 (1966): 7–11.

14. See M. Roche, "Early History of Science in Spanish America," *Science* 194 (1976): 806–10; R. Ferreira, "As origens da atividade científica no Brasil," *Ciência e Cultura* 30 (1978): 1301–1307; M. S. Giambiagi and M. Giambiagi, *Alcune reflessioni suggerite dal tema piani di studio per il dottorato in chimica teorica, IX Congreso de Chimici Teorici di Espresione Latina* (Stresa, Italy, September 1978).

15. See Eric J. Hobsbawn, *Industry and Empire* (London: Penguin), p. 146, as qouted by O. G. Velho, *Capitalismo autoritário e compesinato* (São Paulo: Difel, 1976).

16. J. Leite Lopes, "Science and Dependent Development," *Interciencia* 2 (1977): 138–42.

17. O. G. Velho, *Capitalismo.*

18. F. A. Biato, E. A. De Almeida Guimarães, and M. H. Poppe Figuereiro, *Potencial de pesquisa tecnológica no Brasil* (Brasília: Ministério do Planejamento, Instituto Econômico e Social, 1971), pp. 1–198; J. Leite Lopes, "Les transferts de technologies: l'exemple du Brésil," in *Plurisciences, Encyclopedia Universalis* (Paris, 1978), pp. 221–24.

# 18

# The Need of the Third World for Humanism

## Indumati Parikh

Indian humanists, like humanists all over the world, want to build a cosmopolitan, cooperative commonwealth of men and women based on freedom, knowledge, truth, and secular morality. This is not something new. Men of thought cherished it in practically all civilizations from time immemorial. India had its share in the philosophy of Sankhya, Vaishesikhi, and Lokayat, propagated around 800 B.C. by sages like Kapila, Kanad, Charvak, and others. In a later period, by 600 B.C., Gautama Buddha brought about a social and religious revolution in India by founding a casteless society and a secular religion which spread over not only India but the whole of Southeast Asia. After quite a few centuries, reactionary Brahminical religion overthrew Buddhism from India and established a caste-ridden, ritualistic Hindu religion which is still dominating Indian civilization.

Modern-day humanism wishes to use reason and the methods of science to help resolve human problems, and this is the awesome task of humanists in India.

# I

The Indian subcontinent is one of the unique regions where different cultures have encountered each other and yet no one could claim to have established its own culture throughout the length and breadth of the region. If one studies the Indian subcontinent of today, it is an odd and interesting mixture of cultures or perhaps a number of tribes, religions, and civilizations. This continent is still inhabited by aboriginals as well as the most modern communities, each wanting to preserve its own identity and life style. There is a large majority of people, who believe and follow ancient traditions, a religious way of life, rituals, the caste system, etc., while at the same time they use modern technology without caring to know the scientific truth about it. There seems to be a revival of religious rituals among the modern educated people to establish their identity in the atmosphere of growing religious fundamentalism, perhaps out of fear of alienation.

The real problems of the whole subcontinent are poverty, ignorance, the population explosion, and rising fundamentalism, resulting in social insecurity and ultimately leading to authoritarianism. Most of the governments in the region are paying only lip service to these problems without making real efforts to tackle them. Very few citizens are even aware of what is happening around them, their day-to-day worry being to get two square meals a day and some sort of shelter to protect them from the sun and the rain. The question then arises why people are not revolting, why they think in terms of religion when they are starving, why they are not changing the social or political order. These are some of the questions we have to address before we can think of helping the people to improve their life stance or expect them to join the humanist movement.

The explanation could be found in the fact that for centuries India had accepted a hierarchical social structure—the caste system—with dominance of a priestly class. All other castes were subordinate to Brahmins in every walk of life. To be born in a higher caste was the result of one's fate. One suffered in this life because of sins committed in the previous life. Every action of a person was determined by religion and the caste. One had no right to change these facets. The outcome of this was naturally the fatalistic outlook and slavish mentality. Even today, it is firmly ingrained in the Hindu mind, including that of the so-called educated people. This is perhaps the biggest stumbling block in the path of progress and it is taking a long time to convince people that "Man is the maker of his world."

Is it not natural, then, that nearly forty-five years after independence, the majority of the people are not guaranteed the basic needs for human existence? We are moving in a vicious circle of poverty, ignorance, population explosion, rising fundamentalism, and a fatalistic attitude to life. One has to break it, and we are convinced that only the acceptance of the humanist philosophy will help us to break it. It is undoubtedly a long way, and it is a difficult path to be covered, and yet it is the only way to achieve the goal. Therefore, before teaching any philosophy or perhaps along with its teaching, one has to improve the living conditions of a large mass of humanity. For that, we need to develop their own will and get their cooperation in solving their problems. These people have to forget their caste hierarchy and differences, their religious differences, and accept each other as human beings.

Unfortunately the people in power do not want it. India—the largest nation of the subcontinent—has accepted "democracy" and also "secularism" (if only on paper) as the cornerstones of the political system of governance. The two ideas could be explained to the people and their own small worlds could be organized on these two principles. This has to be the basis of humanist action at the grass root level. Humanists in the Third World should understand that it is the people of these countries who would bring the social change needed to improve their own lot. The ruling powers or the political parties struggling to capture power are not going to help the people. A primary interest of many in power is to keep people ignorant and poor. They want to foster religious fundamentalism so that they can keep their power by suppressing people's liberties. Poor, ignorant people could be incited to religious frenzy and at the time of elections used as a vote bank. All this makes the task of the humanist in the Indian subcontinent arduous. Unfortunately humanists outside the subcontinent perhaps are not fully aware of this situation.

## II

India in a way is slightly better off with respect to individual freedom and human rights. But the other nations on this subcontinent are perhaps worse situated. Two of them are open theocracies and are reviving ancient inhuman laws to govern their people. Liberal progressive thought is suppressed in Pakistan. I can quote from a recent letter I received from a member of an organization, "The Enlightenment," which believes in "equal opportunities to all, girls, boys, men, and women." The person writes:

> We are a humanist-minded secular organization amidst a medieval-style Islamic fundamentalism. I have been declared non-Muslim (excommunicated) for my humanist and secular activities concerning civil and human rights. Social, moral, and economic sanctions are being imposed on me as a part of persecution, as well as on other free thinking individuals. In this religious darkness imposed in the name of Islam and supported by many Christian clergy, we need the support of humanist and secular individuals and organizations from all over the world. We look forward to your cooperation.

A similar letter received from Sri Lanka stated that rationalism and secularism has all but disappeared from Sri Lanka after the death of its most famous rationalist, Dr. Kavour, in 1982. Does the humanist world in the developed countries know the plight of the Third World? Are they ready to understand us and help us? These are some of the questions all of us from the Third World have been asking of ourselves. I raise these questions also for my fellow humanists of the West.

How did this Third World come into existence? In a sense it has been in existence for the last 150 years or more as a result of colonial expansion by the industrialized West. The British who conquered India came to this subcontinent to find cheap raw materials for their expanding industries and markets for their finished products. The European Renaissance had freed the Western people from the dominance of religions and more or less also from supernatural dominance. The Enlightenment and Industrial Revolution were the products of the same.

This led to the emergence of nation-states in Europe which enabled them to consolidate economic and political power leading to empire building. These empires, as we all know, were far more exploitative in nature than the earlier ones. They also led to concentration of power in a few hands and put army and police at the central place of their governance. On the other hand, it led to the division of the world into West and non-West. To my mind, this was the beginning of the emergence of the so-called Third World.

Profit was the principal motive force behind the development of the economic system of the world, marked by the capitalist system. The West would keep its own people happy and affluent because of exploitation, sometimes in human exploitation, in the form of slave markets in their colonies. But they developed breakneck competition among themselves which led to the world wars, one more destructive than the earlier one. The Second World War brought down the mighty empires and a number of free nations came into existence all over the world. These people, who were kept in the backwaters of modern civilization in the interest of the West, were backward economically, politically, educationally, and culturally. Years of slavery, lack of exposure to modern science and technology and to administrative skills naturally confused them and most of them are still grappling to get a firm footing in the modern world after nearly forty to fifty years of independence.

The Indian subcontinent had to undergo a traumatic experience of partition on a religious basis and also communal discord as the aftermath of partition. Several causes of this situation could be found in the native soil, yet nobody can absolve the British of the responsibility of creating conditions which made it impossible for them to leave the country without dividing it. In a way, it was a result of their policy of divide and rule which they had used right from the start and continued to follow, particularly after the 1857 revolt, which was a joint effort of all the communities to overthrow the British rule in India.

India no doubt has benefitted by the contact with the British and the liberal thought that was dominating them at home in the nineteenth century. This introduced the Indian intellectuals to modern science, technology, and also modern political institutions. The Indian elite absorbed many of Britain's democratic ideals and when freedom was given, Indians decided to accept democracy as a way of governing themselves. India has also a better chance of survival and development because of a large land mass, natural resources, and the skeleton administrative structure which was intact when the British left. It also inherited all the industries developed during the British rule and a large number of educated people who had some idea of what was expected of them as free citizens.

# III

When the British left, the Indian subcontinent was a disturbed land. The economy was a jumble of all systems. Its biggest productive segment—agriculture—was in a primitive state. Zamindari or landlordism, a feudal system of land ownership, existed in the Indo-Gangetic Plain. Industries were developed in such a manner that they were meant to supplement the British industrial requirements without any thought to the needs of common people of India. Natural resources were not optimally harnessed. The educational system was not producing clerks and middle level managers needed by the British administrators, though a few professionals were produced by the universities. By and large, common men and women remained illiterate. Technical skills were not developed, nor was much attention paid to science education, the development of the scientific way of thinking, or science research. In addition to this, Pakistan chose to be a theocracy and also a totalitarian regime. This led to further discord between the governments on the two sides, though a large number of Muslims remained in India of their own choice. India had accepted democracy and secularism as its ruling principles from the beginning, to accommodate all the religious groups. But people were not ready for both. Indian society was and is to a large extent a hierarchical society. Caste diversities were sharp and these dominated everyday life of the community. Added to these were the divisions emanating from regional and linguistic diversities, each wanting to preserve its own identity. There was no common language and English, though the language of administration, was spoken or understood only by a small fraction of the population comprised of high government

officials and university-educated people. We had to struggle to keep
our freedom and develop as a democracy in spite of all these difficulties.
If the developed world had helped us, perhaps we would have been
in a much better position. But that world was facing its own problems
and started using the Third World to maintain its balance of power.
The Cold War was fought at the cost of the Third World—playing
one nation against another, establishing army and navy bases in different
newly-formed nations, even fighting their own wars on the soils of these
nations which were used as pawns in the power game. This did not
give us any respite to develop ourselves. India and Pakistan could have
and still could be developed with each other's help and cooperation.
Tod y both the countries are facing a perpetual crisis, and crisis is
described as a situation which is a "turning point when recovery of
death is decisively indicated." We do not know which way we are heading.
We have been groping in the dark more or less for the last fifty years.
Population explosion, poverty, ignorance, and religious fundamentalism
seem to be with us all the time. The people in power or those who
want to come to power are more concerned to remain in power than
to solve the problems. It is in the interest of their power game to keep
a large number of people as they are, because they are afraid of the
masses turning into men and women who could manage their own affairs.
In this situation, we know that humanist philosophy is the only ideology
which will help us. It will cut across our religious, caste, language, and
regional divisions, ultimately cutting across the national divisions also.
We are a very weak and more or less insignificant movement, facing
all odds both on national and international fronts. The international
humanist movement hardly has any idea about our problems. The fault
lies of course with us. We do not make them aware of our situation,
nor perhaps are any of them keen on knowing about us. I am taking
this opportunity to inform you about our problems, and also what the
people in power in your world are doing for us.

## IV

We know that everything is not all right with common men and women in developed countries. There is unrest, there is increasing poverty, unemployment and social insecurity, and increasing ignorance of common people. The rich are getting richer and poor are becoming poorer. False consumerism is encouraged through advertisements and mass media. These two are being used to tranquilize people, instead of making them aware of their problems. The collapse of the U.S.S.R. ended the Cold War and put a break on the proliferation of the armament industry. The situation in the developed world both politically and economically seems like it is leading to a crisis.

And how is the industrialized world facing it? I feel that they are trying to use the same old methods of exploiting the underdeveloped world instead of finding new means to put their house in order. They want to sell arms to people who should not have them, nor can they afford to buy them. Governments unable to meet the most elementary needs of their people are spending money on armaments. Financial institutions like the World Bank and the International Monetary Fund have become institutions which are only recycling the debts of poor countries instead of helping them in their real development. Rich countries are holding India and other countries at ransom because they are refusing to fall in line and change their laws to suit the American industrialists. Why are ecologically dangerous products like Halon 12 H and drugs proved harmful for human consumption dumped in the Third World? Why is there no voice raised in the West against what is popularly known as the intellectual property right being extended to twenty years? Why this embargo on knowledge? Until very recently, no scientist thought of patenting his/her discoveries. It is known that development of science depends on free flow of knowledge. Why make things difficult for the Third World?

I am not asking these questions of any audience, but of the humanist audience, mostly from the Western countries. Why are they not taking cognizance of factors impeding the development of a real global community? All that we read in the Western humanist literature makes us believe that religion is the only stumbling block in establishing a humanist society in the world. Can you develop such a society leaving two-thirds of the world in utter poverty and ignorance and at the mercy of the developed world? I know I am quite harsh in speaking about the developed world, but the people of the Third World are facing an impossible situation both internally and externally. There seems to be no hope for them. We, the humanist groups in this world, would look to you for understanding of our problems and assistance in sorting out these.

I am quite conscious that our people have to take courage in their own hands and face the situation squarely. We need a renaissance—a new type of renaissance which does not take place only at the elite level but at the grass root level in which common people are involved. We have to build up adequate confidence in our people so as to enable them to help themselves, to bring about a social and cultural transformation, change our ways of thinking and of behaving, and understand that fundamentalism would ruin us. A number of grass root organizations, such as women's groups, are springing up and humanists have to be at the forefront of these movements of social change.

What I want to convey to you is that humanists should think of life in a holistic manner. The world cannot be divided into compartments on a regional basis or any other basis. All of us have to understand that we will develop or sink together. No single part of this world can achieve the height of human development leaving a major portion of the world to fend for itself and to sink to the bottom because of forces beyond its control. If humanists are really interested, and I am sure they are, in building up a global community, a cosmopolitan commonwealth of free men and women, they have to change their own common people and particularly the outlook of people in power—political and economic power—and influence them so that the people of the Third World will not be at the mercy of power-hungry people in any country.

We need time and respite, we need peace, we need opportunities to develop ourselves without suppression from the powerful outside world. We need all the available knowledge, for which a free and uninterrupted flow of knowledge is necessary. We do not need arms, nor do we need tranquilizers, but freedom to help ourselves to come into our own without hindrance.

Therefore, I appeal to you all, who are knowledgeable and quite influential in your own world, to understand us, to understand our needs, and to help us to develop. There is no doubt in my mind that the twenty-first century will be the century of humanism, but much depends on the humanists in the developed world and their understanding of the world situation. Many people in the Third World are beginning to understand that acceptance of humanism is their only hope. We have to help the most neglected groups in the world to come into their own. Women and adolescent children should form our target groups for development. Women need education, health, family planning, income of their own, and decision-making power in social, economic, and political affairs. They will have to be made aware of these needs and helped to develop into self-confident citizens. Similarly, adolescent boys and girls will need much help for growing up as responsible and thinking citizens. These are the most urgent tasks of the humanist groups in our world, and this will mean working at the grass-root level and at the same time bringing humanist thought and action to the common men and women as would make them understand it and would enable them to act accordingly. It is an uphill task, but it must be undertaken urgently and on a very large scale.

At the same time we have to do something to influence the policy makers and implementers in our country. They can become a big obstacle along with people in power wanting to remain in power by hook or crook. Therefore, education of voters becomes another urgent and important task for the humanists in countries where some sort of elective processes have been introduced. The voters would have to be made aware that dissemination of power to the people through people's committees, which could act as schools for democracy as well as a powerful weapon in the hands of common people to shape their own lives, could be an effective solution in getting rid of the politicians with a vested interest in power.

The most important and urgent of all the tasks the humanists have to undertake is to get rid of fundamentalism which is being fostered by our politicians. If you meet common people in the street, you will find no trace of hatred or animosity in them against anybody. But the political parties want to divide people so that they could be easily suppressed. Communal differences and hatred are fostered mostly by them. All this is possible for them because of the ignorance and poverty of the masses. Population growth at an alarming rate is also a direct result of these two. Development is the best contraceptive and to understand the value of freedom, and enjoy freedom, some level of cultural development has to be achieved. One cannot hope to raise the cultural level of people who cannot think beyond survival. Minimum human needs must be met, and that is the biggest problem the whole of the Third World is facing today.

The vicious circle has to be broken and I think we have to put our heads, hands, and hearts together and to make every effort to break it. The Third World needs a new type of renaissance, a renaissance that would be a joint effort of a newly emerging intelligentsia, aware of human values, and the common people of their respective communities. These together would rescue abiding values from their ancient heritage, enrich that heritage by whatever is good and human in modern civilization, and form an integral part of the world community of free men and women. We do not need complete leveling of the world. I know there will be different levels of cultural growth, and, in spite of the oneness of the globe, there will be diverse cultures flourishing simultaneously on the globe. All that we want is to live in harmony and peace with each other and grow like human beings, free of all impediments to our growth. Only in this way could we evolve a global ethical society, consisting of free human beings, who are happy in their own diverse worlds, each one enriching the other. The Third World would then merge into the common human world and help its own citizens to develop without fetters or fears.

# 19

# Peace and Progress

## Mourad Wahba

The title of this paper necessitates a clarification of two concepts: peace and progress. Concerning peace we have to distinguish between two meanings. We say that a certain country is currently at peace with its adjacent neighbors. When we speak thus, we are using the word "peace" negatively to mean the absence of actual fighting. In contrast to this negative meaning of the word peace, there is the positive meaning of the word, when we use it to say that peace exists among people living in civil society under civil government; they are at peace with one another, and that is called civil peace based not on the absolute but on the relative, not on dogmas but on civil laws.

But peace today is of a global nature as long as interdependence of states prevails. Thus peace cannot be confined to community peace but to the global community that includes all people on earth.

Concerning the concept of progress, there was nothing in the heritage of antiquity to support this concept. The human pilgrimage was seen as a rise from a sinful fall to ultimate grace. Thus, terrestrial history is an interlude between two unchanging eternities. In this case, to expand the advance of science into a general doctrine of progress would have been out of time with the status quo that was full of myths.

It was with the beginning of the seventeenth century that the notion of progress was known and elaborated. It meant that things will get better in the future and implies also the assumption that a pattern of change exists in the history of mankind. This means that the notion of progress involves a synthesis of the past and a vision of the future.

Within this context, Condorcet in the eighteenth century, the age of Enlightenment, expressed his notion of progress in his book entitled *Sketch for a Historical Picture of the Progress of the Human Mind* (1795), which has often been regarded as the authentic testament of the Enlightenment, written by the last of the *philosophes*. He divided the historical progress of mankind into nine stages, adding a tenth epoch in which he sketched his hopes for the future on the basis that the experience of the past, and the observation of the progress that the sciences and civilization have already made in the analysis of the human mind and the development of its faculties, are the strongest reasons for believing that nature has no limit to the realization of his hopes.[1]

But this concept of the future is of a secular nature, whereas it was of a religious nature in the fifth century according to St. Augustine's concept in his famous book *The City of God,* where he looked forward to the movement of history from life in the earthly city to life in the heavenly city. Such extra-logical preoccupation continued to affect thought and action throughout the subsequent centuries.

Throughout the nineteenth century, side by side with the spirit of progress that animated the minds of the philosophers, there was to be seen slowly developing a kind of malaise affecting the very premises on which the spirit of progress rested. It was a reaction against modernity as manifested in such major phenomena as industrialism, technology, science, secularization, and liberation from traditional values. We get intimations of it from Burke and Durkheim.

Burke's writings and speeches revolve around the following items: reverence for the divine disposition, reliance upon tradition, conviction that men are equal in the sight of God. In the *Reflections on the Revolution in France* (1790), these items generally find their most earnest expression. Society is indeed a contract and the state ought to be considered nothing better than a partnership." As the ends of such a partnership cannot be obtained in many generations, it becomes a partnership not only between those who are living but also those who are dead and those yet to be born. Each contract of each particular state is but a clause in the great primeval contract of eternal society, linking the lower with the higher nature, connecting the visible and invisible world, according to a fixed compact sanctioned by the inviolable oath which holds all physical and all moral natures, each in their appointed place.

Now two questions have to be posed:

1. What is the primeval contract of eternal society?

2. What is the inviolable oath which holds all physical natures in their appropriate places?

I suggest that the contract and the oath were meant to be divinely appointed. Therefore, the French Revolution runs counter to the divine order, and this is the meaning of what Burke called "General Earthquake" in the Western World.

Durkheim, though he described the evolutionary progress of society from a "mechanical" to an "organic" organization, took notice of the tendency for the level of advancement of society to be inversely related to human happiness. Durkheim tells us that modernity brought with it progress but at the same time a heightening of insecurity, and increasingly, of a loss of purpose in living. All this analysis was brought to a focus in Durkheim's study of suicide. In the increase of suicide in the West Durkheim thought that the malaise of an entire society can be seen.

By the beginning of the twentieth century Einstein's theory of relativity surpassed Newton's system and paved the way for nuclear physics, which played a great role in the emergence of nuclear weapons and consequently the threat to the survival of man on earth. A new term was coined by the American philosopher, John Sommerville, that is, "omnicide," meaning the end of humans by humans.[2]

But we have to ask: This omnicide is an outcome of what? It is an outcome either of science or of man. In my opinion, it is an outcome of both. Science is a creative outcome of man defined as a creative animal.

But what is creativity? Creativity is the ability of human reason to establish new relations for the purpose of changing reality. This definition implies two essential concepts: new relations and changing reality. What matters here is that changing reality may be for the best or for the worst. And the criterion of distinguishing between the best and the worst is whether we humanize or dehumanize reality. And I mean by humanizing reality adapting reality in order to cope with the essential needs of the human beings, that is to be conscious of being in unity and peace in this world.

Now let us investigate the history of science to clarify our definition of creativity and its aforementioned consequences. I pick up Einstein's case. In their joint book *The Evolution of Physics* Einstein and Infeld state the following:

A new concept appears in physics, the most important invention since Newton's time: the field. It needed great scientific imagination to realize that it is not the charges nor the particles but the field in the space between charges and the particles. . . . The theory arises from the field problems. The contradictions and inconsistencies of the old theories force us to ascribe new properties in the space-time continuum to the scene of all events in our physical world. The relativity theory develops in two steps. The first leads to what is known as the special theory of relativity, applied only to inertial co-ordinate systems, that is, to systems in which the law of inertia as formulated by Newton is valid. The special theory of relativity is based on two fundamental assumptions: physical laws are the same in all co-ordinate systems, moving uniformly, relative to each other, and the velocity of light always has the same value. From these assumptions fully confirmed by experiment, the properties of moving rods and clocks, their changes in length and rhythm depending on velocity, are deduced. The theory of relativity changes the laws of mechanics. The old laws are invalid if the velocity of the moving particle approaches that of light. The new laws for a moving body as formulated by the relativity theory are splendidly confirmed by experiment. A further consequence of the special theory of relativity is the connection between mass and energy. Mass is energy and energy has mass. The two conservation laws of mass and energy are combined by the relativity theory into one, the conservation law of mass-energy.[3]

From this creative theory emerged the atomic technology that was used to change reality. But in which direction, for peace or for war? If we go fast from 1905 to Ronald Reagan's "Star Wars" speech, we have to ask: Who was backing this project? The Christian fundamentalism headed by Jerry Falwell. And this means that dogmatism is behind the misuse of science.

Who was guilty on the afternoon of August 6, 1945, when an atom bomb was dropped over Hiroshima: the scientist or the dogmatist?

Heisenberg's answer is as follows:

> The word "guilt" does not really apply, even though all of us were links in the causal chain that led to this tragedy. Otto Hahn and all of us have merely played our part in the development of modern science. We know from experience that it can lead to good or to evil. But all of us were convinced, and especially our nineteenth-century rationalist predecessors with their faith in progress, that with growing knowledge good would prevail and evil could be kept under control. The possibility of constructing atom bombs never seriously occurred to anyone before Hahn's discovery; nothing in physics at the time pointed in that direction. To have played a part in so vital a scientific endeavor cannot possibly be considered quilt.[4]

But if the scientist is not guilty who else is guilty? In my opinion the guilty party is the dogmatist who thinks that he grasps the absolute truth and wishes to impose it by force. That dogmatist was the *homo fascistus.* "Believe, obey, fight" was the injunction Mussolini gave to the Italian masses.

Today the dogmatist is the fundamentalist, whether he be Christian or Moslem or Jewish fundamentalist.

To conclude, one may say that the main obstacle against global peace is dogmatism. Creativity according to my definition is its antidote.

# Notes

1. Condorcet, *Selected Writings,* ed. by Keith Michael Baker (Indianapolis: The Bobbs-Merril Company, 1976), p. 259.

2. J. Sommervile, *Nuclear War in Omnicide,* quoted in Michael Allen Fox and Leo Groazke, *Nuclear War* (New York: Peter Lang, 1985).

3. A. Einstein and L. Infeld, *The Evolution of Physics* (Cambridge: Cambridge University Press, 1971), p. 244.

4. W. Heisenberg, *Physics and Beyond* (New York: Harper & Row Publishers, 1971), p. 194.

# 20

# The Terrors of Islam

## Antony Flew

Great and terrible systems of divinity and philosophy lie round about us, which, if true, might drive a wise man mad.

Walter Bagehot, 1879

Before accepting any belief one ought first to follow reason as a guide, for credulity without enquiry is a sure way to deceive oneself.

Celsus, about 170 A.D.

Everyone concerned to promote emancipation and enlightenment has been exhilarated by the historic changes of these last five years: first among the self-styled peoples' democracies and then within the former USSR itself. Spectacular developments within what was until only yesterday the Socialist Bloc came at the end of two decades during which, first in Greece, then the Iberian peninsula, and later generally throughout Latin America, military dictatorships and other authoritarian, but not totalitarian, regimes were replaced by elected governments. All this, along with some signs of similar developments even in Africa, led Francis Fukuyama to ask in a much talked of article (published in *The National Interest* in Summer 1989) whether we are witness to "The

End of History?"

By this, of course, was not meant either an end of all historical events—such as the subsequent Gulf War—or that there will never again be an authoritarian or totalitarian regime anywhere. It was, rather, that such regimes as do survive or emerge will never again be seen as elements in an irresistible wave of the future. And that, from now on, rulers who have not been elected, and—what should be seen as much more important—rulers who cannot in due course be removed through an equally free election, will everywhere be accounted aberrant and radically illegitimate.

## I

There is however another totalitarian ideology which needs to be considered here. For many of its ever more numerous adherents see themselves as militants of a movement of which the eventual worldwide triumph is guaranteed not just by an impersonal hypostatized History but by an omnipotent intending Agent. When in 1920 Bertrand Russell visited the USSR—decades before the Politburo found it convenient to present itself as the Protector of the Arabs—he discerned similarities between Bolshevism and Islam: "Bolshevism combines the characteristics of the French Revolution with those of the rise of Islam";[1] and "Marx has taught that Communism is fatally predestined to come about; this produces a state of mind not unlike that of the early successors of Mahommet."[2] So Russell himself concluded: "Mahommedianism and Bolshevism are practical, social, unspiritual, concerned to win the empire of this world. . . . What Mahommedianism did for the Arabs, Bolshevism may do for the Russians."[3]

As a clear, commendably honest, and altogether authoritative epitome of the totalitarian character of Islam consider this manifesto issued in Leicester, England, on behalf of the Islamic Council of Europe:[4]

The religion of Islam embodies the final and most complete word of God. . . . Departmentalization of life into different watertight compartments, religious and secular, sacred and profane, spiritual and material is ruled out. . . . *Islam is not a religion in the Western understanding of the word.* It is a faith and a way of life, a religion and a social order, a doctrine and a code of conduct, a set of values and principles, and *a social movement to realize them in history.*[5]

In this we have a statement which satisfactorily transcends all differences within and between various Muslim communities, such as those between Sunni and Shi'a or between the so-called Fundamentalists and their opponents. The term "fundamentalist" is anyway in the present case peculiarly inappropriate. It is derived from the title of a series of tracts—*The Fundamentals,* published in the United States in 1909; and it is defined as the belief that The Bible, as the Word of God, is wholly, literally, and infallibly true—a belief that, notoriously, commits fundamentalist Christians to defending the historicity of the accounts of Creation given in the first two chapters of Genesis. To rate as truly a Christian it is by no means necessary to be in this understanding fundamentalist. It is instead fully sufficient whole-heartedly to accept the Apostles' and/ or the Nicene Creed. But in order to be properly accounted a Muslim it is essential to be a fundamentalist with regard to (not the Bible but) the Koran.

The crux is that, whereas only a very small proportion of all the propositions contained in the Christian Bible are presented as statements made directly by God in any of the three persons of the Trinity, the Koran consists entirely and exclusively of what are believed to be divine revelations made through the Prophet Muhammad, and made therefore in Arabic.[6] These revelations are supposed to have been "received, in circumstances of a trance-like nature, over a considerable number of years intermittently, the first . . . dating from about A.D. 610 and the last shortly before Muhammad's death in A.D. 632. . . . Tradition relates that a few years after his death the scattered fragments were collected together." But it was only "during the reign of the third Caliph Uthman (644–656) that the definitive canon was established by a panel of editors directed by the Prophet's amanuensis Zaid ibn Thābit."[7]

As might be expected, given this method of compilation, the resulting book is extremely repetitious. Although, except for the first and one of the shortest, the 114 surahs or chapters are arranged only in descending order of length it would be difficult if not impossible to suggest a better alternative. Unlike the Christian Bible, which contains several sorts of writings—historical essays, collections of psalms and of proverbs, four separate accounts of the preachings and eventual martyrdom of Jesus, letters to branches of his young and growing church, and so on—the Koran is a collection of recorded teachings coming from or through a single mouth. One consequence is that the theologians of Islam have little room to maneuver; either for softening harshnesses which prove embarrassing; or for developing doctrines found or supposedly implicit in some books but not in others. So the history of Islam provides no parallels to the disputes about, for instance, the nature of the Trinity[8] which have riven Christendom.

The Prophet Muhammad appears as the final Messenger from the Mosaic God of Judaism and Christianity. A total of twenty-eight predecessors are named in the Koran as having previously been assigned to spread the message of obedience to Him. One of these, mentioned frequently and always respectfully, is "Jesus Son of Mary." But the crucial Christian claim that Jesus was the Son of God is categorically repudiated, and those who persist in maintaining that claim are condemned to "a painful chastisement":

> They are unbelievers
> who say, God is the Third of Three
> No god is there but
> One God
> If they refrain not from what they say, there
> shall afflict those of them that disbelieve
> a painful chastisement . . .
> . . . and in the chastisement they
> shall dwell forever.[9]

That "painful chastisement" will thus clearly be for the offense of heretical belief, and, as the Koran asserts repeatedly, it will not only consist in the infliction of extremes of agony but also—infinitely more severe—continue eternally. Every surah begins "In the Name of God, the Merciful, the Compassionate," while the first proceeds forthwith, as sooner or later do most of the rest, inconsistently to indicate that there is to be a Day of Doom on which the mercy and the compassion of "The All-Merciful, the All-Compassionate" will be revealed to be strictly and very narrowly restricted:

> Praise belongs to God, the Lord of all Being
> the All-merciful, the All-compassionate,
> the Master of the Day of Doom.
>
> Thee only we serve; of Thee alone we pray for succor.
> Guide us in the straight path,
> the path of those whom Thou hast blessed,
> not of those against whom Thou art wrathful
> nor of those who are astray.

The central, fundamental, continually and emphatically repeated message of the Koran combines a promise with a threat. The promise is to "those who believe, and do deeds of righteousness—theirs shall be forgiveness and generous provision";[10] a generous provision including among other attractions not only "G rdens of Delight" but also "wide-eyed houris as the likeness of hidden pearls."[11] The threat is to unbelievers, "those who strive against Our signs to avoid them—they shall be inhabitants of Hell,"[12] a habitation in which "garments of fire shall be cut for the unbelievers" and where "for them await hooked iron rods as often as they desire in their anguish to come forth from the eternal fire."[13]

Since no attempt is ever made to reconcile these threats of eternal torture for too belatedly repentant unbelievers with the endlessly reiterated assertion that Allah is "the All-Merciful, the All-compassionate" it becomes appropriate to recall a very characteristic observation made by Thomas Hobbes in his *Leviathan:* "In the attributes which we give to God we are not to consider the signification of philosophical truth, but the signification of pious intention, to do him the greatest honor we are able."[14]

The Koran calls for belief and consequent obedience. It is, surely, calculated to inspire fear, indeed abject terror, rather than love. So it is altogether appropriate that the apologetic argument since attributed to and named for Pascal was employed centuries earlier by the famous Sufi theologian al-Ghazzali, who died in A.D. 1111.[15] Allah is presented in the Koran, notwithstanding "the attributes . . . of pious intention," as a cosmic oriental despot who penalizes perceived disobedience and crushes perceived opposition by eternally extended exercises of uninhibited[16] and total power.

The qualification "perceived" has to go in since actually to oppose or to disobey we should need to believe in the existence and orders of the despot, while any who actually knew that opposition or disobedience was to be rewarded by eternal torture and yet chose to disobey or to oppose would, by simply engaging in such egregiously and inordinately insane behavior, show themselves not to be fit and proper subjects of punishment.

The sentences which Allah is to hand down on the Day of Doom are alleged to be just, despite the inordinate disparity between finite offences and infinite penalties, because "every soul earns only its own account; no soul laden bears the load of another."[17] By the hypothesis that is no doubt correct, but only so long as "the load of another" is construed as the load of another human being. Yet the Koran contains an abundance of passages paralleling those notorious hard sayings of St. Paul which insist that God, conceived not only as the omnipotent initiating and sustaining cause of the entire universe and of everything in it but also as punishing some of its creatures, must therefore be recognized to be punishing those creatures for offenses for which, by the hypothesis, God alone cannot escape the ultimate sole responsibility. All such Divine discriminations have therefore to be seen as arbitrary exercises of total power:

> Therefore hath he mercy upon whom he will have mercy and whom he will he hardeneth. Thou wilt say then unto me, "Why doth he yet find fault? For who hath resisted his will?" Nay but, O man, who are thou that replies against God? Shall the thing formed say to him that formed it, "Why has thou made me thus?" Hath not the potter power over the clay, of the same lump to make one vessel unto honor, and another unto dishonor?[18]

The first parallel passage in the Koran comes at the very beginning:

> As for the unbelievers, alike it is to them,
> whether thou has warned them or hast not warned them,
> they do not believe.
> God has set a seal on their hearts and on their hearing,
> and on the eyes is a covering,
> and there awaits them a mighty chastisement.[19]

Christian apologists never cease from arguing that by endowing humankind with the dangerous attibute of free will God breaks the chain of causation and thus escapes responsibility for our actions. A similar attempt was made in the early days of Islam.[20] But the overwhelmingly dominant position is that argued in "A Popular Theological Statement."[21] Reasonably enough, since the fact that we are indeed creatures who can and cannot but make choices, some of which are made of our own free will and others only under various degrees of compulsion or constraint, does not preclude the possibility, and on these assumptions the necessity of Divine causation making us the particular creatures who actually make whatever choices we do in fact freely make in whatever senses we do indeed freely choose to make those choices.[22] Thus this Statement insists that "it is the duty of every Muslim to believe" that "Both good things and evil things are the result of God's decree." Nevertheless it continues, "Modern theologians sometimes teach that God has the duty to be good, to do good for people, to will the good. . . . God has no such duty. If He had . . . His free will and power would be limited, which is clearly contrary to the dogma of omnipotence and the divine will."[23]

The same Statement grasps the key to understanding what it is to have a choice; and why it is impossible for those who have acquired the concept of choice to deny that they are members of a kind of creatures which can, and cannot but, actually make choices: "Any person discovers a difference in moving a hand and when the air moves it."[24] The distinction between these two fundamentally different kinds of bodily movements is, of course, that the former—label these movings—are, while the latter—label those motions—are not, under the direct control of the person, the agent, whose movements they are. But although our movings are thus always and necessarily under our direct control, and although alternative movings or an abstention from moving must therefore always be possible, none of this provides any guarantee that it is not the agency of God which makes us all the different people who choose in the various senses in which we do severally choose. However, if it were, then it would not do to respond by maintaining that, for that reason, we really and truly could not have done other than we actually did. For the crucial expression "could do otherwise" is, surely, itself definable only ostensively and by reference to movings.

## II

It should by now be obvious that Islam is one of those "Great and terrible systems of divinity . . . which, if true, might drive a wise man mad." So are there evidencing reasons for concluding that it is certainly or even very probably true? If we were forced to conclude that it is more probable than any non-hell-threatening alternative, then that conclusion would surely constitute a powerful motivating reason to persuade ourselves of its certain truth.[25]

The question whether Muhammad was a Messenger from the Mosaic God must be distinguished from the logically prior question whether there is such a sender of Messengers. The affirmative answer to that logically prior question the Koran takes absolutely for granted, presupposing in the reader or hearer knowledge of or derived from both the Bible and "some sort of native Arabian tradition."[26] Of the twenty-eight predecessors mentioned most attention is given to Moses and after him to Jesus and, curiously, Noah. Moses too is the one who is alleged to have been supplied with the most spectacular credentials in the shape of the Plagues of Egypt.

To those familiar with traditional Christian apologetic it is remarkable that no claims are made about miracles allegedly worked by or on behalf of Muhammad himself—at any rate if we except the contention that the composition of the Koran, which is apparently agreed by all those competent to judge to be the supreme masterpiece of Arabic literature, itself constitutes a miracle.[27] This omission gave purchase to the objection which Aquinas took to be decisive. Muhammad, Aquinas wrote:

> . . . did not bring forth any signs produced in a supernatural way, by which alone divine inspiration is appropriately evidenced; since a visible action which can only be divine reveals an invisibly inspired teacher of truth. . . . It is thus clear that those who place any faith in his words believe frivolously.[28]

The only contemporary supposed signs to which the Koran appeals as evidences are various familiar facts of nature described as the achievements of Allah. For instance:

> Those are the signs of the Book;
> and that which has been sent down to thee
> from thy Lord is the truth, but most men
> do not believe.
> God is He who raised up the heavens
> without pillars you can see,
> then He sat Himself upon the Throne,
> He subjected the sun and the moon,
> each one running to a term stated,
> He directs the affair. . . .
> It is He who stretched out the earth
> and set therein
> firm mountains and rivers,
> and of every fruit he placed there two kinds,
> covering the day with the night.
> Surely in that there are signs for people who reflect.[29]

And, after some more of the very similar, "surely in that are signs for people who understand." But a reflection which thus proceeds immediately from visible facts to their Invisible Cause, and an understanding which is manifested in this inference, must be prejudiced. The prejudicially drawn conclusion may be correct. For prejudices are not necessarily and as such mistaken. But, absent the prejudicial assumption, such arguments are manifestly unsound. For they proceed directly as, Aristotle might have put it, from actuality to impossibility: from descriptions of what to all appearance occurs normally and naturally to the conclusion that these phenomena cannot really be what they appear to be but are instead the products of supernatural agency.

Some lines from *Uncle Tom's Cabin* are more revealing here than perhaps the authoress recognized. For, unlike the Yankee Miss Ophelia, poor Topsy had never been theologically indoctrinated by either parent or preacher. Yet she had had abundant opportunity to learn from rural observation what in my young day urban fathers used to reveal to schoolbound sons as "the facts of life." So it is Topsy who answers for unprejudiced common sense and common experience:

"Do you know who made you?" "Nobody, as I knows on," said the child with a short laugh. The idea appeared to amuse her considerably; for her eyes twinkled and she added: "I s'pect I grow'd. Dont thin' nobody ever made me."[30]

## Notes

1. *The Practice and Theory of Bolshevism* (London: Allen and Unwin, Second Edition, 1962), p. 7.

2. Ibid, p. 27.

3. Ibid, p. 74. For a critique of this book compare my "Russell's Judgement on Bolshevism," in *Bertrand Russell Memorial Volume,*. ed by George W. Roberts (London: Allen and Unwin, 1979), pp. 428–54.

4. I chose a statement from this organization, rather than anything from a traditionally Muslim country, in order to draw attention to the fact that there has since World War II been a still continuing Muslim immigration into the United Kingdom, France, Germany, and other countries of Western Europe. Both by natural increase and through further immigration all these minorities are growing both absolutely and relative to the non-Muslim majorities. Compare, for instance, Mervyn Hiskett, *Some To Mecca Turn to Pray* (London: Claridge, 1993); also the publications of Majority Rights, B.M. Box 3515, London, WCIN 3XX.

5. Quoted in Whitefield Foy, ed., *Man's Religious Quest* (London: Croom Helm/Open University, 1978), emphasis added.

6. That is why a proper Islamic education has to include the learning of Arabic and the consequent study of the original and only authentic text of the Koran.

7. Quoted from the first paragraph of A. J. Arberry's Introduction to his translation, first published by Allen and Unwin in London in 1955. Subsequent references will be to the Oxford University Press World's Classic paperback edition of 1985, giving the number of the surah in Roman and of the page in Arabic numerals.

8. Any such doctrine, of course, as a defection from austere, absolute, unequivocal monotheism, is utterly repugnant to Islam.

9. V, pp. 112 and 113.

10. XXII, p.339. Nowhere can I find any suggestion that unbelievers who "do deeds of righteousness" might be excused resurrection and hence damnation.

11. LVI, p. 560.

12. XXII, p. 339.

13. XXII, p. 335.

14. Chapter XXXI.

15. Compare "Is Pascal's Wager the Only Safe Bet?" in my *God, Freedom and Immortality* (Buffalo, N.Y.: Prometheus Books, 1984).

16. Uninhibited, that is to say, by any enlightenment objections to torture or other cruel and unusual punishment. The notorious this-worldly punishment of manual amputation is specifically prescribed by the Koran: "And the thief, male and female; cut off the hands of both, as . . . a punishment exemplary from God" (V, p. 106). To reject it as the outrage which it is must therefore be to become less of a Muslim, albeit a somewhat more enlightened human being.

17. VI, p. 142.

18. Romans IX, 18–21.

19. II, p. 2. Compare VI, pp. 123, 125, 134, and 136; VII, pp. 161 and 166; IX, p. 194; XVI, p. 269; XXIV, p. 357; XXXIX, p. 474; and LXXXI, pp. 632–33.

20. See, for instance, Andrew Rippin and Jan Knappert, eds., *Textual Sources for the Study of Islam* (Manchester: Manchester University Press, 1986), pp. 17ff.

21. Ibid, pp. 126–34. I have been assured by one of the editors that "it is found in many Islamic countries. . . . It has been accepted as dogma for a long time among Shafeitic and Hanafitic scholars. Divine predestination and (the absence of) human responsibility are discussed almost verbatim in many

other tracts and treatises in Islamic lands. Islamic theology is homogeneous."

22. For further discussion see Antony Flew and Godfrey Vesey, *Agency and Necessity* (Oxford: Blackwell, 1987). This work quotes Luther's comment upon this necessity: "Now by 'necessity' I do not mean 'compulsorily' a man without the Spirit of God does not do evil against his will, under pressure, as though he were taken by the scruff of his neck and dragged into it, like a thief or a footpad being dragged off against his will to punishment; but he does it spontaneously and voluntarily." It also cites his desperate response to the challenge quoted earlier from Romans: "The highest degree of faith is to believe that He is just, though of his own free will he makes us . . . proper subjects for damnation and seems (in the words of Erasmus) 'to delight in the torments of poor wretches and to be a fitter object for hate than love.' If I could by any means understand how this same God . . . can yet be merciful and just, there would be no need for faith." (pp. 88–89).

23. Rippin and Knappert, *Textual Sources,* p. 133. Islam therefore spares itself the Problem of Evil: the insoluble problem, that is to say, of reconciling the actual existence of evil and the putative existence of a God who is not only all-powerful but also all-good.

24. Ibid, p. 134.

25. For development of such an argument, distinguishing evidencing from motivating reasons for belief, see the article referenced in note 15, above.

26. Andrew Rippin, *Muslims: Their Religious Beliefs and Practices* (London and New York: Routledge, 1990), vol. 1, p. 15.

27. I am reminded of a column in *The Times* of London in which, during the Mozart Bicentenary celebrations, Bernard Levin urged canonization, naming what he judged to be suitably miraculous compositions.

28. *Summa contra Gentiles,* Book I, Chapter 6, Section 4; pp. 73–74: in the first volume of the Doubleday Image edition. For the difficulties of evidencing such naturally impossible occurrences, see my *Atheistic Humanism* (Buffalo, N.Y.: Prometheus Books, 1994), Ch. 3.

29. XIII, p. 239. And compare XLII, p. 502; XLV, p. 516; and LXXVII, p. 642.

30. Harriet Beecher Stowe, *Uncle Tom's Cabin* (New York: Books Inc., undated), p. 206.

# 21

# A Humanist Perspective for Britain*

## Bernard Crick

For a long time a debate has been waged between Christian and secular rulers on the question whether democracy is the product of the Christian faith or of a secular culture. The debate has been inconclusive because, as a matter of history, both Christian and secular forces were involved in establishing the political institutions of democracy; and the cultural resources of modern free societies are jointly furnished by both Christianity and modern secularism. Furthermore there are traditional nondemocratic Christian cultures to the right of free societies which prove that Christian faith does not inevitably yield democratic historical fruits. And there are totalitarian regimes to the left of free societies which prove that secular doctrine can, under certain circumstances, furnish grist for the mills of modern tyrannies. The debate is, in short, inconclusive because the evidence for each position is mixed (R. Niebuhr, *Christian Realism and Political Problems.*

"The evidence for each position is mixed," indeed. Niebuhr famously wished to build bridges between the two positions. One bridge that he offered for "a strong affinity . . . between democracy and Christian-

*Originally published in H. Willmer, ed., *20/20 Visions* (London: SPCK, 1992).

ity" was that "the toleration which democracy requires is difficult to maintain without Christian humility."[1] Perhaps that is one way of putting it; but in this context, "skepticism" is often, if not an exact synonym for humility, at least an acceptable historical substitute. "Modern morality," said the philosopher and social anthropologist Ernest Gellner, "does in fact accord respect to honest doubt, rather than to ill-founded conviction." But his remark also cuts in both directions, restraining Christian and humanist alike. Actually I'm not sure that modern morality does stop short of giving respect even to ill-founded convictions, so long as they are sincerely held (an exaggerated respect for sincerity and authenticity is what many have against modern morality). The breaking point between respect and rejection is not people holding ideas of any kind, but when they try to impose them on others.

I raise this theme of the ambiguity of tolerance by way of prelude; I will expand it later. But let me begin at the beginning, before Niebuhr's democracy and before the Christian revelation even, with the Greeks and the Romans. We need to consider the nature of politics itself, both philosophically and as something with a cultural history, therefore an origin; and an origin, both as speculation and as practice, no older than the Greek city states.

## The Nature of Politics

By politics, then, I mean exactly what Aristotle meant. It is an activity among free men living as citizens in a state or *polis;* how they govern themselves by public debate. Of course, even in Aristotle's great book (or rather lectures) *The Politics,* the word was used for any type of government as well as in this special sense. And the special sense to him was not necessarily, at any given time, democratic. A *polis* must have a democratic element in it, but he favored mixed-government, the able rotating and governing with the consent of the majority (and even that excluded slaves, foreigners, and women, of course). A pure democracy, he said, would embody the fallacy that because men are equal in some things, they are equal in all. However, the special sense of polis or civic state was to him a conditional teleological ideal: both a standard and a goal to which all states would naturally move if not impeded, as well they might be, by folly, unrestrained greed, or power-hunger by leaders lacking civic sense, or by conquest—which finally settled the matter (as Machiavelli was to notice).

Aristotle brings out the intense specificity of the political relationship (and I will soon say its inherent secularity) when, in the second book of *The Politics,* he examines and criticizes schemes for ideal states. He says that his teacher Plato made the mistake in *The Republic* of trying to reduce everything in the *polis* to an ideal unity; rather, it is the case that

> there is a point at which a *polis,* by advancing in unity, will cease to be a *polis;* there is another point, short of that, at which it may still remain a *polis,* but will nonetheless come near to losing its essence, and will thus be a worse *polis.* It is as if you were to turn harmony into mere unison, or to reduce a theme to a single beat. The truth is that the *polis* is an aggregate of many members.

Politics arises then, according to Aristotle, in organized societies that recognize themselves to be an aggregate of many members, not a single tribe, religion, interest, or even tradition. That is why in my *In Defence of Politics* I defined politics as the activity by which the differing interests and values that exist in any complex society are conciliated.[2]

Politics arises, then, from a perception of differences as natural. But this perception has both an empirical and an ethical component. The empirical component is a generalization that all advanced, complex, or even (just say) large societies contain a diversity of interests—whether moral, social, or economic; and in fact, usually blendings of each, hard to disentangle. The ethical component is that there are always limits beyond which a government should not go in attempting to enforce consensus or unity. Perhaps no limits can be demonstrated in general. They may all be specific to time and place. But the principle of *limitations* is general and the empirical distinction is usually clear, allowing for deceit, rhetoric, and muddle, between regimes that strive to limit power and thus govern politically, and those regimes whose rulers strive after total or at least unchallengeable power. That my definition of politics, or rather Aristotle's, is not an empty truism can be seen at once if one sadly remarks that most regimes even in the modern world are not political: they hunt down politics, not encouraging it as a civic cult; they act politically only when faced with a superior, immovable, or uncertain rival power.

Some call themselves "realists" and say that politics is basically only and all about these differences of interest, a matter of "conflict." Hard-nosed political scientists might accept at least half of St. Augustine's analysis: that any justice in the earthly city is simply self-interest. States hold together for the same reason that bands of robbers hold together: self-love and mutual interest. Others call themselves, or more often are called, "idealists" and say that politics is basically about doing what is right: "where there is no vision the people perish" or "let justice be done though the heavens fall" (as is most surprisingly written over the main door of the Old Bailey). But beware of the fallacy of the excluded middle. It is possible to reject both: "realism" for not allowing enough to at least occasional altruism and sociability; and "idealism" for being prone to dangerous chimeras of human perfectibility. One does not have to be a Christian to hold a tender skepticism (or humility?) about human perfectibility. So a third school says that political morality is about reaching some consensus or agreement about civic procedures, about the institutional conditions of peace and justice, not about the nature of peace and justice themselves: political institutions should build a ring and hold it fairly in which all comers can debate and attempt to get their way (well, nearly all comers; not those who try to smash up the ring—as even John Stuart Mill agreed). As a *politique,* I am obviously of this third school.

You will notice that I have moved from talking of morality in politics to "political morality." I mean that it is the mark of politics that it often has to conciliate, in some manner that is both right and acceptable, rival codes of morality as well as material interests. This is close to what Kant meant by "practical reason," or what Max Weber called an "ethic of responsibility" rather than "absolute ethics." The first advocate of toleration as state policy, Jean Bodin, argued that it was not the business of the state to punish heretics, but only to keep the peace; but he also believed that God would damn them hereafter.

## Religion and Politics

If in considering the future in Britain, politics involves dilemmas for the religious, religions can also pose dilemmas for the political. A little later I must remind us that the whole concept of toleration (which is a curious concept anyway) arose no earlier than the seventeenth century in reaction to wars of religion. And it would be bad taste to dwell on the fact that these wars were not metaphorical and were among Christians. In the last twenty years I have attended numerous conciliation conferences or meetings, both open and clandestine, in or about Northern Ireland. In the fervor and fury of the new ecumenicity, a Catholic priest and a Presbyterian minister will commonly chant in unison, "As we were the root and cause of this conflict, it is now for us to solve it together." When I was bold enough to air an opinion, I would welcome the first part of the statement as historically more or less true and as removing some inhibitions in discussion, as well as for its possible therapeutic value to those who uttered it; but would then caution that the second proposition was at best a half-truth and, at worst, dangerous hubris. The gentler way of putting this is to say, as I know many Christians do, that one should do one's best, but not take on impossible burdens in society, in the *civitas terrena*. Many of you may feel that the *civitas dei* is entitled to a little spare energy and study.

The point I wish to make is that political activity, whatever its motivation, is a secular activity: the worse for that, said Augustine, only part of the fallen and transitory state of man carried away by the waters of Babylon; good enough, said Thomas Aquinas, part of natural law open to the reason of all mankind, Christian, Jew, and infidel alike, even though, of course, it was incomplete without divine law and the bending of God in grace. Calvinists often said that the sphere of statecraft as such was something "morally indifferent," although whether magistrates acted in a Christian manner was crucial. Christ's injunction was both broad and clear: "Render unto Caesar that which is Caesar's and unto God that which is God's." The principle is clear, even if the application makes it always what philosophers now call "an essentially contestable concept." Nonetheless, it rules certain things out and thereby makes other things plain. Theocracy is ruled out, except as a heresy—however much images of the nature of God affect all our perceptions of the nature and purposes of political authority.[3] Christian theology, unlike some other world religions, is essentially dualistic: some things are one, some the other. They are not of equal value, but there is a division. Put in the simplest terms: Caesar must not dictate to God's servants and God's servants have no business dictating to Caesar on secular matters, at least not in God's name. As citizens they may do what they wish within the limits of Christian belief; but belief is neither a sufficient nor a sure guide in all political action. Unlike the authority of God, the authority of neither church nor state are comprehensive, they are specific. And looked at from the point of view of individuals, the good life, the life of virtue, is no longer political or public life itself as in Greece and Republican Rome: it is a life of prayer, praise, striving for salvation and ministry, even though ministry also has duties toward works and the world stemming from the Beatitudes.

All I point out, from outside, and forgive me this banality, is that such good Christian injunctions as to help the poor, to love our neighbors and to honor the peacemakers, which would, indeed, have a totally transformative and revolutionary effect on political order if everyone acted according to them (which is unlikely both in common sense and by Christian views on human imperfection), these injunctions cannot be forced on people. They cannot be forced both in historical senses—we've never seen it work, only the bloodshed and the horror of forced conversions; and because in philosophical senses there is freedom of will, there is conscience and individuality. Minimally, if moral progress is to be made in removing or modifying legislation or behavior unacceptable to Christians, this progress has to be through politics: convincing and compromising with unbelievers. Maximally, if any still wish to make progress toward a Christian commonwealth, the path must still be through persuasion; unless, as the more eschatological and less cautious of liberation theologians have sometimes suggested in South America and Southern Africa, there is sudden simultaneous mass conversion and changing of hearts. Well, perhaps. There is some ground for this in Christian tradition. But I don't think it reasonable or morally responsible to bank on it.[4]

Even in what seemed until very recently a quite desperate situation, South Africa, the preaching of mass transformation and revolution has now given way to political bargaining and the making of compromises. But, I must in all honesty say, many—perhaps most— of those involved are motivated by Christian principles (or painful rethinking in the Afrikaans seminaries). Yet it would be equally fair to say of apartheid—as Machiavelli said of Philip of Macedon, "moving whole populations as if they were sheep and not men"—that "one does not have to be a Christian to see that that was evil."

It seems to me that it is very important for any group who wish to be taken seriously, be they a party or a church, an ideology or a sect, not to claim too much, always to shave close with great Occam's razor lest in claiming too much one imperils or renders incredible or ridiculous what is truly important. I am irritated, almost angered at times, by that shallow argument that without religion there can be no morality. The question arises, what religion? Or have a limited class of religions a common morality? Some may have some elements of a common theology, but that's a different question. And they all believe in salvation, more or less: views about the ontological status of non-believers do vary a bit. But (and I'll return to this point) it seems to me that if one gives to different religions, to different creeds of Christianity indeed, the respect deserved by the importance of their claims and the sincerity of their claimants, then one must see that different moral codes are involved. Think only of prescriptions on sex and the family—let alone punishment, property, inheritance, diet, and drink.

It would be a long argument, but I am convinced that we all owe to each and every other person, simply in the capacity of their being human, a respect and recognition to another unique and equal individual. This is essentially H. L. A. Hart's view of what is involved in the idea of human rights—Hart a skeptic, differing greatly in grounding but not so much in conclusion from Immanuel Kant's categorical imperative; Kant a Christian, but by faith: morality he held must be a universal derivable from common reason.[5] The good metaphor that we are all brothers and sisters does not depend for its force on there being a father, or knowing who that father is or what his will is. If you lose your faith in the Creator very suddenly, I would still feel safe in your company at dinner and would not expect you to assume you'd be robbed by me because I am not a Christian. I am not a Christian. I lack faith. But intellectually I have always understood, as once I felt when younger, that Christianity is about matters even more important.

I do not hesitate to call things good or bad when moral choices or judgments are called for. Humanists often do themselves less than justice by trying to avoid what social scientists quaintly call "value judgments." I make them all the time; too easily sometimes. This drives some secularists either into a rigid utilitarianism, often trying to count up (or pretending to) unreflective public opinion on complex issues, or into a kind of pseudo-empirical rhetoric. This can come from both Left and Right. "If something is not done about unemployment, society will break down"; or if you take the *Daily Telegraph* rather than the *Guardian,* then substitute the word "crime." What they mean to say is that unemployment and crime are morally wrong; no way to treat people, no way for people to act. A few years ago even Lord Longford was arguing perfectly seriously that the spread of pornography would lead to the breakdown of British society. This was most unlikely. Even he got trapped in empiricist rhetoric. He surely meant that it was wrong, morally repugnant. On the whole, I thought so too. And I argued that if more secular liberals publicly said that porn was rubbish or sometimes intolerably offensive to—or actually threatening to—women, if there was more condemnation or mockery, there might have been less demand for legislation. Because something is morally wrong it does not follow that it should be banned.[6]

However, let me quickly say that a rational secular morality is either an affair of philosophers and intellectuals, demonstrating its possibility, or else of the kind of "innate decency" (Orwell's words) that one still hopes to find in most ordinary working people. There is more sociability or mutual care among working-class neighbors than among the middle classes: capitalist society is like that. To judge by editorials even in the quality press, the old sense of a reasonably clear civic and secular morality is a declining faculty in the middle ground of the opinion makers or facilitators. It lacks institutions and organizations. There was once a feeling that the Labor Party embodied the old nonconformist conscience (which Max Beerbohm said "makes cowards of us all"). But that is an even longer generation ago than when the Conservative Party was openly and unashamedly paternalistic. Both the parties now seem scared of speaking with any other language but that of economic self-interest. And material interests are, remember, indeed half of the political equation but only half: values *and* interests, interests *and* values. So it seems to me that the churches, for reasons quite unconnected with theology, probably far less connected with personalities than is thought, and hardly deliberately, got drawn into what I hope (I nearly say pray) is this temporary vacuum.

The report *Faith in the City* has to be seen in this light. It would never have had much notice or caused such a storm in press and Parliament if the Labor Party had not seemingly abandoned its good old moral rhetoric (out of fear that every time it said anything moral people would feel the taxman's hand in their pocket). They yielded the moral high ground and descended to the battlefield offered by their opponents: that of who will give or take more from whom. In such a decline of genuine political thinking, almost by default the churches began to sound like an effective political opposition. It was churchmen who stated the obvious moral case for the social justice of retaining a steeply graduated income tax—though that is a matter of common morality, not specifically Christian. Hence the counterattack from newly theologized noisy MPs, and the equal danger to the true role of the church of, on the one hand, John Selwyn Gummer's and Enoch Powell's obsessive "Keep Out: no works, only faith" or, on the other hand, of Frank Field's conviction that the Labor Party still marches in a mysterious way toward the New Jerusalem and the Kingdom.


## Toleration and Pluralism

I was taught at the London School of Economics by Harold Laski, an agnostic Jew who preached a socialist version of philosophical pluralism explicitly based on the Anglican divine J. N. Figgis's *Churches in the Modern State*. For churches, we knew that Laski was substituting trade unions when he intoned, "a free church in a free state."[7]

But that analogy in political terms, and this must include the churches when they are involved in political issues, is not a bad one. And now not just a plurality of Christian churches in Britain, but of religions: now Hinduism and Islam as well as Judaism. The internal tensions created by the terrible war in Iraq and the Rushdie affair are still in all our minds. And religion and ethnicity unhappily but quite obviously, if contingently, go together. And if you who are living in England now have to start thinking of Britain as a pluralistic society, religiously, nonreligiously, and ethnically, you had better also start remembering the Scots, before by insult or neglect they begin acting somewhat like the northern Irish, whom we would willingly forget if we could. The Welsh are another question again, having an intense sense of common identity, but either two identities overlapping or one with a formidable linguistic division. Now in a sense this was always so, for three centuries at least. The real question is why the English did not recognize it and, at least in popular consciousness, celebrate the pleasures of variety in a state composed of four nations, not one. Life might have been easier for the new immigrants if their cultural diversity could have been seen as not raising in principle any essentially new problems of administration and popular tolerance, but ones analogous to those of the four nations and the once open hostilities between the Christian churches. Why should the immigrants be asked to be English? They may reasonably be required to be British. The one is a cultural identity, the other is a political and legal set of allegiances. The Scots are Scottish and British, not Scottish and English. But until the English have sorted out this distinction (in which the Anglican community should have a lot of experience to offer), the immigrants are in confusion.[8]

Reacting against the religious disputes of the seventeenth century, toleration both as a state policy and an educated attitude began to spread in Britain in the eighteenth century. But let us remember one essential thing about tolerance. It arises because people do differ on fundamental and important things, but wish to limit the practical effect of their differences. Tolerance is not complete acceptance, still less permissiveness; it is modified disapproval.

If you will forgive a humanist for saying so, I think ecumenicity can be taken too far. I was once privileged to hear the late General

Eisenhower invoking "our common Graeco-Romano, Christian-Jewish tradition": and nowadays his speech-writers might think to mention Black Muslims if not all of Islam. That is just what "The Trimmer" John Savile, Lord Halifax, called—thinking of a pulpit of his day—"good stout resolute nonsense." Theologians today can do better. I wonder? Or are they just better behaved toward each other, which God knows is something. If I understand anything of the meaning of many doctrines of the Catholic Church and the somewhat less precise, which is also to say less dogmatic and inflexible, beliefs of the Anglican Communion, it seems to me beyond reason for theologians to hope to resolve or compromise such differences in committee—except either by negotiated surrender or by the somewhat unpredictable course of what I gather some call progressive revelation. Surely what is much more important and plausible (both morally and politically) than seeking to synthesize dogmas is to gain greater understanding and respect by ordinary church members, not least in Ulster, of each other's *differing* beliefs. At least myth and fantasy might be dispelled. Toleration does need at least some cognitive element. I mean that if one doesn't know what offends others, much accidental offense can result when strongly held creeds coexist in close geographical proximity. Oiling the muskets with pig grease (if that did have anything to do with the Indian Mutiny —a lot of the facts we were taught were wrong facts) shows symbolically at least how specific and unexpected things can get. Try offering a meal to an orthodox Jew. Try reasoning with him about the dietary laws of Leviticus. Why bother? Learn and respect his odd different ways so as to plan intelligently how to deal with meetings. Of course, some reciprocity and mutual empathy helps. This is not always forthcoming from fundamentalists (the word "fanatic" seems out of favor). And there are perhaps some limits to how long it is wise to go on turning the other cheek.

Toleration does not mean that we should search for the highest common factor or the least common multiple of all respectable beliefs, and end up with a kind of entromorphic ecumenical Muslim-Christian-Hindu-Baptist-Holistic unitarianism. We should argue for our rival views of truth and morality vigorously but, yes, tolerantly. And always remember that it is skeptics and opponents we need to convince; preaching to the already saved is a largely useless pleasure. Britain has to become a more and more pluralistic society in every way. This does not mean watering down. It means a greater appreciation of differences and variety. Naturally in any society there are limits to toleration. Not all religious practices are acceptable to other groups, still less to majority opinion. But these limits have to be discussed politically: the will and the example must be there to reach creative compromises about behavior, of course, but never about belief.

## Inconclusions

It seems to me pointless and needless for my fellow British humanists to rail against the truth of religious beliefs, only against the abuse of authority (although I admit that the questions of religious broadcasting and of disestablishment do not "warm my blood likewise," as the Greek anthologist put it, nor seem to me of the highest priority; though in another place I would vote on these issues in the way they have a right to expect of an honorary associate). It seems to me more important in these needlessly cruel and growingly materialistic times to assert the virtues of an active and participant common citizenship and civic morality, the modern form of the Aristotelian politics of creative compromise, and to cultivate a politics of moral judgments as well as of material interests working with any who think similarly. And we do this not as an ultimate end in itself, but as a necessary condition in which free lives and good lives can be cultivated.

Orwell, an agnostic, once said that socialists should not claim to be perfectionist, "perhaps not even hedonistic":

> Socialists don't claim to be able to make the world perfect: they claim to be able to make it better. Any thinking Socialist will concede to the Catholic that when economic injustice has been righted, the fundamental problem of man's place in the universe will still remain. But what the socialist does claim is that the problem cannot be dealt with while the average human being's problems are necessarily economic.[9]

From what I have seen of conciliation groups in Northern Ireland, of university settlements, of members of charities and voluntary bodies working in inner cities among the poor, the destitute, the mentally handicapped decanted to "the community," drug addicts, or sufferers from AIDS, Christians are much more in evidence than my version of the silent majority: Orwell's ordinary people who have no faith but have a sense of common decency, fraternity, sociability, fellow-feeling, call it what you will. This, of course, is only an impression; there certainly are old-fashioned public-spirited humanists in this work, still vigorously counteracting the residual tendency of the Salvation Army to trade soup for souls. And there is the new breed of young professionals with Social Administration diplomas, often a puzzling mix of positive altruistic motivations and negative determination against jobs in business. I don't know whether Christians are more evident in the hard-edge of voluntary work because they are Christians, or simply because they are better organized. To me it doesn't matter. They don't question that one is just working for one's fellow men. I don't question in practice, only intellectually, what the priorities of churchmen should be.

None of these questions is easy. Niebuhr asked, in the extract at the beginning of this chapter, whether democracy owed more to Christianity or to secularism. He answered his own question inadequately when he said that mixed evidence produces an inconclusive debate. We are not dealing here with a debate that could or should lead us to choose between Christianity and secularism as the source of democracy. Both are necessarily and plainly present in democracy, which arises from their interaction. The dualism of Christianity allowed the fame and memory of the citizen tradition of the Greek *polis* and the Roman republic to survive. Civic republicanism was not always particularly honored by Christians until long after the Reformation, but it was not always suppressed. It was the ideal definition of a secularity, the realm of Caesar, that did not challenge (or need not) the realm of Christ. The coincidence of the Renaissance and the Reformation gave these ideas of the scholars and humanists a wider circulation and an unexpected relevance. They were not, of course, democratic. Niebuhr forgot that. Aristotle thought that inherently only some *men* were fit to be citizens, and even the seventeenth-century republicans, while they saw no necessary limitation, advocated or assumed formidable barriers of education—which involved leisure and therefore the possession of property and income. It was when republicanism and Protestantism became involved with each other that the Christian belief in the spiritual equality of man (not equality in much else at the time, heaven knows) took on a political dimension. The first person actually to argue that every human being has a right to be a citizen simply by virtue of being human, not even educated—indeed, better to be simple, uneducated, and natural—was Jean Jacques Rosseau, hating and hated by the Catholic Church but deeply influenced by Protestantism, a kind of crypto-proto unitarian.

A friend of mine who is, I must admit, an Anglican priest, seems to have got the balance and the tension about right. This is a long passage, but it has an interesting twist:

> The church [of England] does not have answers to all the problems the world sets itself. Its proper role is not to aim at being relevant . . . nor to conciliate on all occasions, nor to "influence society." Its role is to proclaim the judgment, the justice, the love of God, and to cooperate with Him in the transformation of this world. In this respect the concerns of the church are otherworldly. The church should not reflect current values and trends, but exists to question and challenge them. Rather than attempting to answer current political questions, Christians might profitably contest the assumptions made by the questioner and examine the terms in which the question is posed. . . .
>
> Church leaders do not have some privileged access to political and social realities. . . . Church leaders and synods should not feel obliged to make a statement on every issue of public concern. The misguided notion that they are so obliged leads to the half-baked, mealy-mouthed and ambivalent character of many ecclesiastical pronouncements. It is tempting to say therefore that church leaders should speak only of general principles and basic values, eschewing at all costs the particular. For two reasons this position is unsatisfactory. First, Christian moral judgment is made initially in the concrete case. The adequacy of a principle is assessed on the basis of the particular actions it entails. Secondly, bishops can talk till they are blue in the

face about general principles, but with no discernible effect. It is when they speak of particulars—the miners' strike, economic sanctions for South Africa, the forcible repatriation of the boat-people—that people sit up and listen.[10]

Dr. David Nicholls starts with a theological point, the priority of witnesses; then points out that bishops have no special authority in politics, but ends, startlingly to some, by advocating highly specific interventions. Yes, politics is like that: principles have to be applied to be meaningful. And the church is not a trade union, but it is as much part of the world as, indeed, a trade union. That is not, I recognize, its primary role. But the interventions of the churches are unavoidable, and often have some special merit; yet political acts even of churches are to be judged by political and secular standards. Politics cannot be divorced from reality; but it is the very *existence* (not the truth) of different moralities that is part of the origin of politics and part of its continual mediating process. If I read Niebuhr's version of Augustine right, there are two worlds to be judged by different standards; but the church is in each of them. It is part of modern society and nobody should pretend otherwise, whether canons of Christ Church, former prime ministers, or my fellow humanists. Recently, the churches have been drawn more into politics than they might intend by the politicians forsaking moral discourse entirely for a discourse of interest, and then savaging all those who seek even to restore a traditional balance. But I believe that for limited but important purposes there is much common ground to work upon between Christians and humanists if the future of British society is not to become as bad as one might rationally fear.

# Notes

1. R. Niebuhr, *Christian Realism and Political Problems* (London: Faber, 1954), pp. 94 and 101.

2. B. R. Crick, *In Defence of Politics,* 2nd ed. (Harmondsworth: Penguin, 1982).

3. D. Nicholls, *Deity and Domination: Images of God and the State*

(London: Routledge, 1989).

4. J. J. Degenaar, "Philosophical Roots of Nationalism," in *Church and Nationalism in South Africa,* ed. by T. Sundermeier (Johannesburg: Raven Press, 1975).

5. H. L. A. Hart, *The Concept of Law* (Oxford: Clarendon Press, 1961).

6. B. R. Crick, *Crime, Rape and Gin: Reflections on Contemporary Attitudes to Violence, Pornography and Addiction* (London: Pemberton, 1974).

7. H. J. Laski, *Studies in the Problem of Sovereignty* (London: Oxford University Press, 1917).

8. B. Parekh, "Britain and the Social Logic of Pluralism," in *Britain: A Plural Society* (London: Commission for Racial Equality, 1990); and B. R. Crick, "Englishness and Britishness," in *National Identities and the Constitution,* ed. by B. R. Crick (Oxford: Basil Blackwell and the *Political Quarterly,* 1991).

9. G. Orwell, in a *Tribune* "As I Please" column, reprinted in *The Collected Essays, Journalism and Letters of George Orwell,* vol. 3 (London: Secker and Warburg, 1968), p. 64.

10. D. Nicholls, "Politics and the Church of England," *Political Quarterly* (April–June 1990): 141–2. This is a special number on "The Political Revival of Religion: Fundamentalists and Others."

# 22

# The Difficulties of
# Today's Religious Apologists*

## George A. Wells

Many Christian theologians now recognize that the situation which has
arisen since the onset of biblical criticism during the Enlightenment (at
the end of the eighteenth century) demands of Christians what one of
their number has called "a different relationship to the Bible from any
appropriate or possible before." According to Christian theologian Dennis
Nineham, "Most theologians would . . . agree that no fully appropriate
relationship has yet been discovered or defined."[1] His colleague Leslie
Houlden voices the same frank and honest uncertainty when he declares.
"Modern N.T. [New Testament] study . . . has made the N.T. no longer
usable in ways long established in Christian devotion," so that it must
be considered as "not . . . determinative for present belief."[2] These theologians are well aware that sayings and deeds have come to be ascribed
to personages in both Old and New Testaments as the outcome of a
long process of editing and transmission, so that the distinction the
Protestants of the Reformation period insisted on, between scripture

*Originally published in G. A. Wells, *Language, Magic, and Religion* (Chicago: Open
Court Publishing), pp. 160–72. Reprinted with permission.

as binding and tradition as not, can no longer be sustained: scripture itself is the outcome of variegated tradition.

Let me briefly illustrate some relevant N.T. examples. Paul's views are reworked by later epistle writers within the canon, writing under his name. Pauline teaching (or what the writer understood as such) is controverted in the letter of James, is declared in the second epistle of Peter to be "hard to understand" and in fact to have been grossly misunderstood. The Acts of the Apostles presents Paul in a very different light from that in which he appears in his own writings. The gospel of Mark is adapted in one way by Matthew, and in another by Luke. The thought of the Johannine Church,. as represented in the fourth gospel, is significantly different from what it is in the epistles ascribed to John; and the fourth gospel diverges very substantially from all three of the others. There are considerable differences concerning the nature as well as the doctrines of Jesus. What the Christian's attitude to the Jew should be—rejection or qualified acceptance—is a matter to which different canonical passages given different answers. Whether one should renounce family ties or cultivate them also depends on which passages one consults, as does what one's attitude to possession of wealth should be, and how far one should extend love to those of other persuasions. Leslie Houlden sums this up with: "N.T. ethics varies from writer to writer,"[3] quite apart from the fact that Christian tradition gives little or no guidance on many important questions.

In many cases, canonical writers were themselves aware that they were opposing the doctrines of other Christians—even, as already indicated, doctrines that are now included in the canon. Theologian John Fenton has shown that "24 out of the 27 N.T. books are to varying degrees the result of controversy among Christians;" that is to say, "89% of the N.T. is the result of Christian disagreement," which was sometimes "extremely bitter." Candor—admirable as it is—is not Fenton's only concern in saying this. He is painfully aware that discord is as rife today among Christians as ever it was in the past; and in pointing to the discord in the earliest documents, he can cheer himself with the thought that, although "we are certainly no better" than our fathers, we are not noticeably worse.[4]

Maurice Wiles, Canon of Christ Church, Oxford, agrees with Fenton that "diversity and conflict . . . has been with Christianity from its inception," and likewise puts some positive evaluation on this fact, saying: "If the truth by which we are to live is not authoritatively given in the past, but continually to be discovered in the present, such a process of discovery is bound to involve experimentation, with attendant error and conflict." It follows, he says, that Christians should no longer regard scripture as "a binding authority," and the Church should no longer give definitive rulings on Christ's virgin birth or bodily resurrection. Wiles says that his critics may object that he is placing scripture under the tyranny of the scholar, who has done so much to raise awkward questions about its value. To this criticism. Wiles relies, appositely, that "what the scholar does is only a more concentrated form of what is involved in any reflective reading of scripture—and it cannot be the church's goal to exclude that."[5]

Theologians who honestly admit to uncertainty in this way are now so numerous among those within Christianity who have real and detailed acquaintance with the Bible that simple discounting of their voices is inappropriate. I am not saying that the relevant issues are to be settled by their—or anyone else's—authority. But when a substantial number of the most serious investigators reach conclusions inimicable to their Holy Orders and/or their professional positions, their arguments are not to be ignored. It is no arbitrary wantonness, but years of study that have driven them to their uncertainty. In this connection, Nineham compares his own experience to that of his tutor R. H. Lightfoot. Lightfoot, he says, was "a traditionalist by temperament if ever there was one," but he was nevertheless "forced by years of gospel study to the conclusion that 'the form of the earthly no less than that of the heavenly Christ is for the most part hidden from us.' " Nineham adds that, for the authors of the N.T.,

> edification was a value in writing about the past at least as important as accuracy, which was in any case impossible for them, at any rate to anything like the degree to which we demand it today. As they saw it, to have written an account of the past which did not conform with the religious beliefs of their time would have seemed irresponsible, whatever the available evidence might be. Indeed lack of conformity with

current religious belief would have seemed to them a sure sign of falsity in any report, however strong its external attestation might be.[6]

The attitude Nineham here specifies is so common—and not only in religious matters—as to be almost ineradicable. In the 1930s, facts discrediting Soviet Russia were widely rejected on the Left as Fascist lies, and facts discrediting Hitler were widely scorned on the Right as Communist inventions. We all have a framework onto which we fit experiences that come our way, and few of us are prepared seriously to question its composition. People acquire a strong prejudice in favor of any view, however cheerful or however gloomy, which they themselves have originated or defended. It may have been adopted reluctantly, but once someone has associated it with himself, with his own reputation, he gets a special liking for it that has no connection with its intrinsic merits.

The great difficulty for the critical theologian is that he cannot just cast the Bible aside. Although Maurice Wiles, as we saw, thinks it is no longer "a binding authority" he still calls it "an indispensable resource."[7] Similarly, Leslie Houlden, although he argues for a certain "provisionality" in religious beliefs, and does not want tentativeness to be impaired by what he calls "sticking points," such as insistence on accepting the resurrection or the trinity—although he is to this extent liberal, he is aware that Christianity must have *some* christological doctrine if it is not to become mere theism. "To be Christian," he says, "is to accord to Jesus unique significance in our relationship with God." How this is to be done, he does not pretend to say. In the latest book of his that I have seen—published last year and entitled *Jesus*—he seems to qualify this commitment to Jesus's uniqueness. For although he insists that he is a Christian, he allows that "the more Jesus is seen in the confines of his historical setting, the more difficult it is to make claims for his significance for all persons of all times and places." And he adds: "There is no evading Jesus as riddle as well as challenge."[8]

John Bowden, an Anglican priest who has pressed his own well informed uncertainty upon the clergy, has found that they responded not so much by impugning his arguments, but by saying: "That's all very well, but what about those of us who have to preach?"[9] People who go to church do not want to be told that nearly all the traditional doctrines don't hold up. As David Edwards, another theologian, has said: they "want religion to be a rock, not a crumbling sandcastle"; they "do not want to be walking question marks following in the steps of a crucified enigma."[10] This makes it very hard for any clergy to face up to the real difficulties when addressing their congregations.

Help in overcoming these difficulties has been sought from certain theories of literary criticism, according to which the text of any literary classic, including the Bible, can include a great multiplicity of meanings. If this is so, we may choose those which we find convenient. John Polkinghorne will be known to you as a professor of theoretical physics at Cambridge who relinquished his chair to be ordained, and is now back at Cambridge as President of Queen's College. He holds that the Bible "must be acknowledged as being polysemous, having multi-layered meanings capable of mediating many messages to its readers." He allows that some biblical passages are morally offensive if taken in their plain sense; there are what he calls "distressing harshnesses in tales of genocide and stoning to death and in the vindictiveness of the cursing psalms." Even the N.T., to which, in the manner of Christian commentators, he is noticeably kinder, raises, he admits, problems because of what he calls its "contrasting points of view" and its at times "disturbingly 'mythical' language."[11] (He puts the word "mythical" in quotes: they are meant as protective, as a signal that "mythical" is not to be understood purely negatively). And so, to avoid being upset or disturbed by a biblical text, we need not, he holds, restrict its meaning to what its author intended it to mean. He calls the Bible "the supreme Christian classic" in the sense that any literary classic—he instances Shakespeare and the Greek tragedians—"has something fresh to say to each inquirer." I find this commonly held view somewhat suspect, and I would suggest that one reason why literary critics have come to hold that classics may ever be interpreted anew is that they have to convince themselves and their readers that there must be something more for them to do than merely to repeat the insights of their predecessors. Polkinghorne allows that what some commentators have claimed to find both in the gospels and in Shakespeare has been "unduly imaginative." He has no sympathy with what he calls "allegorical extravagances," and says there must be no "wilful imposition" of meaning, but "sympathetic exploration" of the biblical text—under, of course, "the guidance of the Spirit" which will guard against aberration. How he can be sure of this I do not know, especially as he concedes that the Bible "is not just there for our detached perusal, but as the vehicle for a personal encounter demanding response."

The kind of instruction theology students receive from such premisses is well illustrated by Jens Glebe-Möller, Professor of Dogmatics at the University of Copenhagen, who confesses himself unable to believe in a God who lets his son die for the sake of humans and who therefore reinterprets traditional christologies so as to make them actual and relevant. For instance, in chapter 5 of Mark's gospel, Jesus cures a woman who had suffered from hemorrhages for twelve years. The professor comments: menstrual bleeding represents sexuality, sexuality is related to violence, and so Jesus's action "seeks to put an end to violence."[12] In this type of exegesis, as in dreams, the slightest resemblance between two things suffices to link them. One hears today talk of "creative accounting," meaning "fiddling with the books." One might, in a similar sense, speak, of "creative interpretation" of ancient texts.

As I have already indicated, it is a whole generation of literary critics who are responsible for the basic idea underlying all this—namely that it is unimportant, sometimes even impossible, to find out what an author meant to say. If this is really so, and it is sufficient to extract mental nourishment from an author's words by the aid of one's own ingenuity, then it is not necessary that the author should have meant to say anything. It follows that authors will be encouraged to write what has no meaning, and critics and students to comment on what has no meaning. Hence the state of affairs prevailing in much literature and literary criticism today. I can claim sufficient experience of literary studies in some of our University Departments to speak with some confidence on this.

Some apologists have gone as far as to claim that we can believe the Bible even if there are no realities to which what it says refers. This view is based on some of the very odd ideas about language current among literary critics. Northrop Frye is a Professor of Literature who has taught at half a dozen universities and holds some thirty honorary degrees. This is what he says on the subject:

> The events the Bible describes are what some scholars call language
> events, brought to us only through words; and it is the words themselves
> that have the authority, not the events they describe. The Bible means
> literally just what it says, but it can mean it only without primary
> reference to a correspondence of what it says to something outside
> what it says.[13]

Of course, if we do not need to ask whether what scripture alleges
to have occurred really occurred at all, then we can bypass some very
awkward questions. Take, for instance, the fierce anti-Semitism of so
much of the N.T. In Acts, orthodox Jews are represented as continually
harassing and persecuting Christians. In the fourth gospel (8:44) Jesus
is made to say that the Jews have the devil as Father. In Matthew
(ch. 23) he mounts a savage attack on scribes and pharisees, and in
the passion narrative of this gospel the Jewish crowd cries out for Jesus's
blood to be "on us and on our children" (27:25). In all four gospels
Pilate appears as a kindly governor whose efforts to save Jesus's life
are thwarted by Jewish malice. Critical theologians are well aware that
all this is propaganda, not history, and that it reflects the interests of
Christian communities of the late first century (when the gospels were
written) in rivalry with Jewish interests. How, then, can something be
saved for the faith from such narratives? This question is discussed by
R. P. Carroll, who has taught the Bible at Glasgow University for more
than twenty years. I must acknowledge that he writes very frankly, saying
for instance: "If reading the Bible does *not* raise profound problems
for you as a modern reader, then check with your doctor and enquire
about the symptoms of brain death." He notes that what he calls the
"anti-homosexual brigade" appeal readily to Leviticus ch. 20, but he
asks: how about the prescription of ch. 19 that no one is to wear a
garment of cloth made of two kinds of stuff? Ought we really to be
checking the labels on our Marks and Spencer's clothing? And if we
ignore this prohibition, why accept the authority of the same text
concerning others? Carroll is clearly aware of the problems. Apropos
of the anti-Semitism of the N.T., he suggests that modern readers might
find it inoffensive if the Jews there are understood as simply symbolizing
rivalry and opposition, as mere "textual Jews," so that the texts refer

to nothing outside themselves, and are not related to history. But in the end, Carroll acknowledges that this really will not do; that "modern readings of ancient texts cannot escape history quite as easily."[14] Similarly, the O.T. scholar B. S. Childs, although attracted to stopping what he calls "the abuses of a crude theory of historical referentiality which has dominated Biblical Studies since the Enlightehment," nevertheless realizes that Christians "have always understood that we are saved not by the biblical text, but by the life, death and resurrection of Jesus Christ who entered into the world of time and space."[15]

So far I have spoken only of difficulties posed by the Bible for theologians. But there are also difficulties raised by the nature of life itself: in particular the age-old question: how can a supposedly omnipotent and omniscient God allow so much undeserved suffering? Again and again, I find apologists admitting themselves baffled here. John Polking-horne's statement of the issue is particularly frank and forceful. He says:

> When we have subtracted all the great load of suffering which arises from man's inhumanity to man, there remains much which does not seem remotely to be our responsibility. To whose charge is to be laid the severely handicapped child who will never have a normal life? The thirty-year-old dying of cancer with half a life unfulfilled? The elderly person whose life ends in the prolonged indignity of senile dementia? If there is a God, surely these things must be his responsibility. It seems that either he who was thought of as the ground of the moral law is not himself wholly good, or he is opposed by other equal and conflicting powers in the world. Either way, the Christian understanding of God would lie in ruins.
>
> I believe that this problem of theodicy, of understanding God's ways in the light of the mixture of goodness and terror which we find in the world, constitutes the greatest difficulty that people have in accepting a theistic view of reality. For those of us who stand within the Christian tradition, it remains a deep and disturbing mystery, nagging within us, of which we can never be unaware.[16]

It used to be widely believed that undeserved suffering in this life would be compensated in the next, and there are passages in the N.T. which support such a view. But decline in belief in an after-life has, for many, closed this solution to the problem. The principal alternative Christian response to undeserved suffering has been that God's care of the world is beyond man's limited comprehension, and that what is apparently cruel is simply to be accepted as an element in this care. If so, then the believer in a loving God does not have different expectations concerning the facts and realities of life than the atheist. The difference is then that the believer has a different *attitude* toward these facts: and some theologians have even inferred from this that the word "God," properly understood, refers not to any objective reality, but only to certain attitudes. Such theologians have been termed instrumentalists, rather than realists. Theological realists insist that there are divine realities, whereas the instrumentalists believe that religious language provides symbols which guide the believer, but are not to be taken as making reference to a cosmos-transcending being. Don Cupitt is an obvious example. He, an Anglican priest, believes that "God must be internalized," "withdrawn from external reality and as it were sucked into the indiviual subject." "God is needed, but as a myth."[17]

Apart from this problem of undeserved suffering, there is the additional fact that any doctrine which allots man a unique place in the universe is hard to reconcile with a zoological conception of the human race. Neither morality nor thinking is peculiar to man. A dog can be as courageous, as affectionate, and as self-sacrificing as any human being. Of course, man's thoughts are more elaborate and play a more important part in his behavior than those of other animals. Nevertheless, every mammal gets to know its environment well enough for its needs. Man's scientific knowledge is more extensive but not otherwise different from the gorilla's knowledge of his forest and its other inhabitants. Man's enhanced brain has made him come to look before and after his present situation to a much greater extent than any other mammal; and one consequence of looking into the less immediate future and guiding behavior by distant expectations is a willingness to endure discomfort and labor for the sake of some anticipated good. Any animal, if he is to learn to behave in a particular way, must be rewarded whenever he does so until the habit is acquired. But unlike dogs, human beings will often be satisfied with a promise, and it is only necessary that they should feel confident that it will be honored. The dog must get his biscuit regularly if he is to learn the trick that brings his master to supply it. It is no good giving a dog a special treat on Tuesday because he behaved himself exceptionally well on Monday. He will not appreciate the connection. Man, however, has only to be persuaded that his reward for good behavior will eventually be forthcoming. In such a case his adaptation to his present situation is a *mental* adaptation. He has not altered his environment to suit his needs, but has merely convinced himself that in the long run, all will be well. In fact, man lives in a world created to some extent by his own brain, and his happiness depends on the kind of world he creates there. The world as he believes it to be is more important to him than the world as it really is, provided only that his belief is not in danger of being shaken by events; and that will depend on a number of factors, among which how often and how severely the belief is tested by experience is an important, but not necessarily decisive one. Disconfirming evidence has been known to result in an obstinately deepened conviction, especially if the believer is not isolated but a member of a group who can support each other. Hence there are Christians who still believe that Jesus is coming soon. Admittedly, disconfirming evidence may eventually become so overwhelming that a belief collapses. But in many cases, clear and decisive evidence is simply not forthcoming. Someone could not long continue to believe that he could live on air, or that it never rained on Thursdays; but where beliefs concern the remote past or the remote future, there is little by way of direct experience to prevent people from believing almost anything. And religions—Judaism, Christianity, and Islam are obvious examples—commonly guide behavior by reference both to a remote past and to a remote future.

In terms of man's two possible forms of adaptation—altering the situation or altering belief—we may say that on the whole science aims at altering the environment, whereas religion—and much philosophy, too—aim at altering belief. The scientist enables us to some extent to alter the environment, sometimes to our advantage, so that we escape adversity, while the theologian and the philosopher teach us ito endure it. ("Liberation" theology is, in this regard, quite exceptional.) They expound the history of our own country, or party, or church in such a way as to comfort or even flatter us, and they depict our future prospects in the same way. I was quite surprised to find religious apologists frankly admitting that much doctrine in religion is of this kind. For instance, the late Bishop and Professor of Divinity. R. P. C. Hanson, having asked "why should we believe the word witnessed to by the Bible and taught by the church to be true?" replied: "What more attractive and satisfying account can be found of our destiny, purpose and status in the world? . . . What makes believers believe is the attractiveness of the proposition which the Christian faith presents to us." He did not, of course, maintain that the only criterion for the truth of a theory is the pleasure to be derived from it: but he does say that, "if we start from this point," we shall find other kinds of authority for the Christian faith "falling into place."[18] This is a good illustration of the way speculative thinking can pass from the service of practical needs to the preservation of inward harmony and contentment. Other animals have little use for such thinking, for what is within their perceptual range requires prompt action, and what is outside it causes them neither comfort nor misgivings. So we find no trace of religion or philosophy among other mammals: they live for the moment and the immediate future. But man has become aware of many lurking perils and needs protection from his fears. As the perils are often imaginary, they may he averted by imaginary measures; and the mental processes which effectively serve this important purpose are to a considerable extent free from disconfirmation by the control of events. Hence it comes that man, the cleverest of all the animals, is the most prone to fantastic errors.

Even here, however, the religious apologists face difficulties. It is hard not to be aware of *some* facts which militate against acceptance of comforting theories. I have already instanced the problem of reconciling the vast amount of undeserved suffering with belief in a kindly God who has overall control. Again, although it might be nice to attribute much or all of the hatefulness and misery of the world to ungodliness, it is hard not to be aware of how often evil has been worked by religious persons. J. H. Mahaffy, the well-known Irish classical scholar who died in 1919, tried to represent the culture of ancient Greece as mere culture of the intellect, without the moral forces to balance it that are supplied by Christianity. Yet he had to admit that "the Italian states in the Renaissance of the 15th and 16th centuries were torn by all the vices and crimes which Thucydides describes as rife in the warring Greek republics" and that, in his own day, there is so much greed and atrocity that "one might be tempted to say that the teaching of Christianity has made, alas! but little difference."[19] The whole passage shows an eminent historian struggling to maintain a strong prejudice in spite of blatant facts of history.

The more sophisticated can avoid any confrontation with countervailing facts by constructing attractive-sounding beliefs that are too vague to collide with reality. I don't propose to go into this technique here, beyond noting the late J. A. T. Robinson's *Honest to God* as a signal example of it. The less sophisticated apologists often simply ignore any evidence compromising their beliefs. As Leslie Houlden has said: "untenability is rarely a mortal disease in theology, and often the patient never notices it." While a learned and candid theologian today tries to come to terms with what he calls "the crude and embarrassing problem of the gospel miracles,"[20] a recent report by the National Church of England Society for Promoting Religious Education[21] justifies the evangelizing of children by finding Jesus's proximity to them reliably documented in John 6:9, where the boy supplying the loaves and fishes was obviously close at hand when the 5,000 were fed. The same report argues that children can "grasp the concept" that "doing what Jesus says makes for a better way of life." There is no indication here that a better way of life can result from implementing only a very selective compilation of the precepts ascribed to Jesus in the gospels, and that such selectivity would have to be justified. Nor is there any awareness of the difficulties biblical scholars have encountered in achieving any such justification. Nineham speaks for many of them when he observes: "if . . . we are asked to pattern our life and beliefs on the teaching and attitude of Jesus, it is not at all clear that we have sufficient evidence to be sure what his attitude or teaching was."[22]

The biblical criticism that has led to all this questioning of the Bible's authority has greatly increased our understanding of Judaism and early Christianity. But is there any loss that goes with this gain? Discarding traditional beliefs can have disruptive effects. Gilbert Murray said that anthropology seems to show that "inherited conglomerates" (his term for tradition, tribal custom, taboo, and superstitions) "have practically no chance of being true or even sensible; and, on the other hand, that no society can exist without them or even submit to any drastic correction of them without social danger."[23] Approval of one's fellows and fear of censure are among society's chief integrating forces. But if they are to be effective, there must exist a climate of opinion, a common attitude of approval and disapproval toward certain forms of behavior. If this attitude is not universal or nearly so, its influence on the individual is much diminished. In a community where it is permissible to put everything into question, the climate of opinion becomes unsettled, and certain acts are no longer universally approved or disapproved, so that the ordinary person's main motive for social behavior is very much weakened.

I would like to conclude by suggesting that there is a problem here for the atheist as well as for the Christian. A society cannot exist for long if its members are always concerned exclusively for their own immediate interests. Their behavior must be adapted to social conditions, and human instincts have been developed in accordance with this necessity, in that, while we certainly possess instincts which prompt us to defend our own interests against rivals, we also have instincts which prompt us to protect some at least of our associates. However, even if we ignore the special promptings of instinct in our theorizing on ethical matters, we can see that it is in our interest that there should be certain general rules of behavior within our community to protect us from injury or exploitation. We must also see that, if these rules are to be generally adhered to, every individual must respect them, and therefore we ourselves must respect them. Nevertheless, it has to be admitted that there are circumstances in which it is appropriate to break them. "Thou shalt not kill" is a good rule, but was it wrong of Hitler's generals to try, in 1944, to blow him to pieces? In sum: no rules at all would make life in society totally insecure; completely inflexible rules will lead to much unnecessary suffering; yet how can one allow exceptions without breaking down the consensus on which stability depends?

## Notes

1. D. E. Nineham, *The Use and Abuse of the Bible* (London: Macmillan, 1976), pp. 234f.; Response to D. L. Edwards, in the latter's *Tradition and Truth* (London: Hodder and Stoughton, 1989), pp. 297, 300.

2. J. L. Houlden, *Connections: The Integration of Theology and Faith* (London: SCM, 1986), pp. 24, 35f., 56, 94, 132, 159, 164f.; *Jesus: A Question of Identity* (London: SPCK, 1992), pp. 1, 101, 118.

3. Ibid.

4. J. C. Fenton, "Controversy in the New Testament," *Studia Biblica 1978*, vol. 3, ed. E. A. Livingstone (Sheffield: JSOT, 1980), pp. 97–110.

5. M. Wiles, "Miracles in the Early Church," in *Miracles,* ed. C. F. D. Moule (London: Mowbrey, 1965), p. 225; "Scriptural Authority and Theological Construction," in *Scriptural Authority and Narrative Interpretation,* ed. Garrett Green (Philadelphia: Fortress, 1987), pp. 51–55.

6. See references in note 1.

7. See references in note 5.

8. See references in note 2.

9. J. Bowden, *Jesus: The Unanswered Questions* (London: SCM, 1988), p. 207.

10. D. L. Edwards, *Tradition and Truth* (London: Hodder and Stoughton, 1989), p. 256.

11. J. Polkinghorne, *The Way the World Is* (London: Triangle, 1983), pp. 20–21, 46, 50–51; *Reason and Reality* (London: SPCK, 1991), pp. 64, 66ff.

12. J. Glebe-Möller, *Jesus and Theology: Critique of a Tradition,* Eng. trans. (Minneapolis: Fortress, 1989), pp. 9, 26, 40.

13. N. Frye, *The Great Code: The Bible and Literature* (London: Ark Paperbacks, 1983), pp. 60–61.

14. R. P. Carroll, *Wolf in the Sheepfold: The Bible as a Problem for Christianity* (London: SPCK, 1991), pp. 2, 81, 101f.

15. B. S. Childs, *The New Testament as Canon* (London: SCM, 1984), p. 545.

16. See references in note 11.

17. Don Cupitt, *Taking Leave of God* (London: SCM, 1980), pp. 8, 39, 120.

18. R. P. C. Hanson, "The Authority of the Christian Faith," in *Theology and Change,* ed. R. H. Preston (London: SCM, 1975), pp. 124–25.

19. J. H. Mahaffy, *Survey of Greek Civilization* (London: Macmillan, 1897), pp. vi–vii.

20. See references in note 2.

21. *All God's Children? Children's Evangelism in Crisis: A Report from the General Synod Board of Education and Board of Mission* (London: National Society [Church of England] for Promoting Religious Education, 1991), pp. 37, 54.

22. See references in note 1.

23. G. Murray, *Greek Studies* (Oxford: Clarendon, 1946), p. 67.

# Appendix

# Members of the Academy of Humanism

*Secretariat*
**Co-secretaries:**

Paul Kurtz, professor emeritus of philosophy, SUNY at Buffalo, editor of *Free Inquiry*
Vern Bullough, former dean of natural and social sciences and distinguished professor emeritus, SUNY College at Buffalo
Antony Flew, professor emeritus of philosophy, Reading University
Gerald Larue, professor emeritus of archaeology and biblical studies, University of Southern California at Los Angeles
Jean-Claude Pecker, professor emeritus of astrophysics, College de France; member of the Academie des Sciences and Europaea Academia

*Executive Director:* Timothy J. Madigan

## Humanist Laureates

*George O. Abell, professor of astronomy, University of California at Los Angeles (USA)
Pieter Admiraal, physician (Netherlands)

---

*Deceased

Steve Allen, author, humorist (USA)

Ruben Ardila, psychologist, National University of Colombia (Colombia)

*Isaac Asimov, author (USA)

*Sir Alfred J. Ayer, fellow of Wolfson College, Oxford University (UK)

Kurt Baier, professor of philosophy, University of Pittsburgh (USA)

Dame R. Nita Barrow, ambassador to the United Nations from Barbados (Barbados)

Sir Isaiah Berlin, professor of philosophy, Oxford University (UK)

*Brand Blanshard, professor emeritus of philosophy, Yale University (USA)

Sir Hermann Bondi, professor of applied mathematics, King's College, University of London; Fellow of the Royal Society; Past Master of Churchill College, London (UK)

Elena Bonner, the widow of the late Academy member Andrei Sakharov and a noted defender of human rights in Russia and the former Soviet Union

Bonnie Bullough, former dean of nursing, State University of New York at Buffalo (USA)

Vern Bullough, former dean of natural and social sciences and distinguished professor emeritus, State University of New York College at Buffalo (USA)

Mario Bunge, Frothingham professor of foundations and philosophy of science, McGill University (Canada)

Jean-Pierre Changeux, College de France, Institute Pasteur, Academie des Sciences (France)

Patricia Smith Churchland, professor of philosophy, University of California at San Diego; adjunct professor, Salk Institute for Biological Studies (USA/Canada)

Bernard Crick, professor of politics, Birkbeck College, University of London (UK)

Francis Crick, Nobel Laureate in Physiology, Salk Institute (USA)

Richard Dawkins, Fellow of the New College, Oxford University

José Delgado, professor and chairperson of the Department of Neuropsychiatry, University of Madrid (Spain)

Milovan Djilas, author, former vice-president of Yugoslavia (Yugoslavia)

Jean Dommanget, Royal Observatory of Belgium (Belgium)

Paul Edwards, professor of philosophy, Brooklyn College (USA)

Luc Ferry, professor of philosophy at the Sorbonne and the University of Caen (France)

Sir Raymond Firth, professor emeritus of anthropology, University of London (UK)

*Joseph Fletcher, professor emeritus of medical ethics, University of Virginia Medical School (USA)

Antony Flew, professor emeritus of philosophy, Reading University (UK)

Betty Friedan, author, founder of the National Organization for Woman (NOW) (USA)

Yves Galifret, professor emeritus of neurophysiology at the University P. and M. Curie; general secretary of l'Union Rationaliste (France)

Johan Galtung, professor of sociology, University of Oslo (Norway)

Murray Gell-Mann, Nobel Laureate; professor of physics, California Institute of Technology (USA)

Stephen Jay Gould, Museum of Comparative Zoology, Harvard University (USA)

Adolf Grünbaum, professor of philosophy, University of Pittsburgh (USA)

Herbert Hauptman, Nobel Laureate; professor of biophysical science, State University of New York at Buffalo (USA)

Alberto Hildago Tuñón, president of the Sociedad Asturiana de Filosofia, Oviedo (Spain)

*Sidney Hook, professor emeritus of philosophy, New York University (USA)

Donald Johanson, Institute of Human Origins (discoverer of "Lucy") (USA)

Sergei Kapitza, physicist at the Institute for Physical Problems, Moscow, and Chair of Physics at the Moscow Institute of Physics and Technology (Russia)

George Klein, author, humanitarian, and cancer researcher at the Department of Tumor Biology, Karolinska Institute in Stockholm (Sweden)

*Lawrence Kohlberg, professor of psychology, Harvard University (USA)

György Konrád, novelist, sociologist; co-founder, Hungarian Humanist Association (Hungary)

Paul Kurtz, professor emeritus of philosophy, State University of New York at Buffalo (USA)

Gerald A. Larue, professor emeritus of archaeology and biblical studies, University of Southern California at Los Angeles (USA)

Thelma Lavine, Clarence J. Robinson professor of philosophy, George Mason University, and president of the Society for the Advancement of American Philosophy (USA)

José Leite Lopes, director, Centro Brasileiro de Pesquisas Fisicas (Brazil)

*Franco Lombardi, professor of philosophy, University of Rome (Italy)

Iolé Lombardi, organizer of the New University for the Third Age (Italy)

Andre Lwoff, Nobel Laureate in physiology and medicine; professor emeritus of microbiology, Institute Pasteur (France)

Paul MacCready, Kremer prize winner for aeronautical achievements; president, AeroViroment, Inc. (USA)

Mihailo Markovic, professor of philosophy, University of Belgrade (Yugoslavia)

Adam Michnik, historian, political writer, and co-founder of KOR (Workers' Defense Committee), the organization of dissident intellectuals that helped to end communist rule in Poland (Poland)

*Ernest Nagel, professor emeritus of philosophy, Columbia University (USA)

*George Olincy, The Andrew Norman Foundation (USA)

Indumati Parikh, president, Radical Humanist Association of India (India)

John Passmore, professor of philosophy, Australian National University (Australia)

Octavio Paz, Nobel Laureate in Literature, author (Mexico)

Jean-Claude Pecker, professor emeritus of astrophysics, College de France, Academie des Sciences (France)

*Chaim Perelman, professor of philosophy, University of Brussels (Belgium)

Wardell Baxter Pomeroy, psychotherapist, author (USA)

Sir Karl Popper, professor emeritus of logic and scientific method, University of London (UK)

W. V. Quine, professor emeritus of philosophy, Harvard University (USA)

Marcel Roche, permanent delegate to UNESCO from Venezuela (Venezuela)

Max Rood, professor of law; former Minister of Justice (Netherlands)

Richard Rorty, professor of philosophy, University of Virginia (USA)

Carl Sagan, professor of astronomy, Cornell University (USA)

*Andrei Sakharov, professor of physics, University of Lomorossov in Moscow, Nobel Peace Prize winner (USSR)

Léopold Sédar Senghor, former president of Senegal; member of the Academie Française (Senegal)

Wole Soyinka, Nobel Laureate, playwright (Nigeria)

Svetozar Stojanovic, professor of philosophy, University of Belgrade (Yugoslavia)

Thomas S. Szasz, professor of psychiatry, State University of New York Medical School, Syracuse (USA)

V. M. Tarkunde, chairman, Indian Radical Humanist Association (India)

Richard Taylor, professor emeritus of philosophy, University of Rochester (USA)

Rob Tielman, copresident, International Humanist and Ethical Union (Netherlands)

Peter Ustinov, actor, director, writer, and humanitarian

Kurt Vonnegut, Jr., novelist (USA)

Mourad Wahba, professor of education, University of Ain Shams, Cairo; president of the Afro-Asian Philosophical Association (Egypt)

George A. Wells, professor of German, Birkbeck College, University of London (UK)

Edward O. Wilson, professor of sociobiology, Harvard University (USA)

*Lady Barbara Wootton, former Deputy Speaker, House of Lords (UK)